CYBERSE
OF INDUSTRIAL INTERNET
OF THINGS (IIoT)

While the Industrial Internet of Things (IIoT) continues to redefine industrial infrastructure, the need for thriving cybersecurity measures has never been more pressing. *Cybersecurity of Industrial Internet of Things (IIoT)* contends with the critical question of how to secure IIoT systems against the intensifying risks posed by cyber threats and attacks. This book is a fundamental guide for industrial leaders and professionals pursuing to understand and implement effective cybersecurity solutions. It covers the fundamentals of cyber-physical systems (CPS), Internet of Things (IoT), and IT systems, while handing insights into prevailing and emerging cyber threats targeting industrial environments.

This thorough resource connects theory and practice, providing a reasoned theoretical foundation upon practical knowledge. Each chapter builds a vital link between academic research and industry application, making it a valuable tool for both cybersecurity professionals and industrial practitioners. With a focus on the latest hacking techniques, multidisciplinary standards, and regulatory frameworks, this book prepares readers with the skills and knowledge needed to protect their industrial infrastructure now and into the future.

Key Features:

- Detailed exploration of industrial infrastructure architecture and communication protocols.
- Understanding of traditional security methods and the threats facing IIoT systems.
- Extensive analysis of IIoT vulnerabilities and effective countermeasures.
- Inspection of reference frameworks, standards, and regulatory requirements for industrial cybersecurity.
- Advisement on cybersecurity risk assessment methods and the implementation of protective measures.

This book is thoughtful for cybersecurity professionals working in cyber-physical systems and critical infrastructure domains, including smart cities, aerospace, and manufacturing. It is also a valuable resource for Chief Information Officers (CIOs), industrial engineers, and researchers engaged in industrial engineering. Whether you are a practitioner, a professional, or a student casting to upskill, *Cybersecurity of Industrial Internet of Things (IIoT)* provides the essential tools and insights to navigate the emerging perspective of industrial cybersecurity.

CYBERSECURITY OF INDUSTRIAL INTERNET OF THINGS (IIoT)

Atdhe Buja

CRC Press
Taylor & Francis Group
Boca Raton London New York

CRC Press is an imprint of the
Taylor & Francis Group, an **informa** business

Designed cover image: Shutterstock Image Id 1398093008

First edition published 2026
by CRC Press
2385 NW Executive Center Drive, Suite 320, Boca Raton FL 33431

and by CRC Press
4 Park Square, Milton Park, Abingdon, Oxon, OX14 4RN

CRC Press is an imprint of Taylor & Francis Group, LLC

© 2026 Atdhe Buja

British Library Cataloguing-in-Publication Data
A catalogue record for this book is available from the British Library

ISBN: 978-1-032-46781-8 (hbk)
ISBN: 978-1-032-46783-2 (pbk)
ISBN: 978-1-003-38325-3 (ebk)

DOI: 10.1201/9781003383253

Typeset in Caslon
by SPi Technologies India Pvt Ltd (Straive)

Dedicated to my family Donika (wife) and Oda and Jeta (daughters)

Contents

VIII   Contents

Author Biography

Atdhe Buja, Ph.D., Commonwealth University of Pennsylvania Bloomsburg. Atdhe Buja is an assistant professor in the Department of Computer Science, Math, and Digital Forensics at the Commonwealth University of Pennsylvania Bloomsburg. He is a world-renowned cybersecurity expert with decades of experience and a leader in information technology and Industrial IoT Cybersecurity. His work has been presented at several conferences, FIRST Cybersecurity, Hackathons, Astana IT University, Balkan Cybersecurity DCAF, Japan, Cybersecurity Workshops, Safer Internet Day, etc. He has participated many times in DEFCON, is a well-known expert on CERT teams where he developed and built CERT in academia and private (ICT Academy CERT), and has held a variety of roles in the cybersecurity industry. He has developed several video courses for the industry and the ICT Academy on VAPT, Incident Handling, SOC, Pentest, Machine Learning Introduction, IoT Security, Attack Scenarios and Incident Response, and Cyber Systems and Network Forensics. He has led many research projects for the ICT Academy with Astana International Scientific Complex and Global Cyber Alliance, developing innovative solutions leveraging machine learning, artificial intelligence, and cybersecurity to meet the needs of wireless sensor networks, the Internet of Things, and Industrial IoT.

His research work focused on cybersecurity countermeasures for Industrial IoT infrastructures and wireless sensor networks (WSNs). He holds a patent on those fields, given by the Government of the Republic of Kazakhstan. He currently teaches computer science and digital forensics at the Commonwealth University of Pennsylvania Bloomsburg, United States of America. He is passionate about helping organizations protect their critical infrastructure from cyberattacks. His latest work can be found on his professional website: https://www.atdheb.com/. You can also follow him at @atdhebuja on X.com.

Preface

The prompt advancement of technology has significantly transformed industrial settings, establishing the Industrial Internet of Things (IIoT) as a basic component of Industry 4.0. This book delves into crucial elements of cybersecurity in IIoT systems, bringing a meticulous guide for understanding, protecting, and enhancing industrial infrastructures.

The chapters start by laying the foundation and introducing the components of industrial infrastructure along with the communication architectures that support them. Traditional security approaches are revisited to highlight their relevance in contemporary contexts. The book progresses to address emerging threats and vulnerabilities specific to IIoT, offering insights into advanced attack vectors and mitigation strategies.

Chapters deep dive into cutting-edge methodologies, frameworks, and countermeasures, introducing models for securing IIoT systems. Each chapter connects theory with practical applications, ensuring applicability to real-world industrial challenges. The focus on governance, incident handling, and the cybersecurity lifecycle underscores the need for structured, adaptive strategies.

This book aims to equip professionals, researchers, and students with the knowledge to protect IIoT settings, fostering innovation while mitigating risks. Arching over foundational principles and modern approaches provides a valuable resource for securing the future of industrial systems.

1

INTRODUCTION TO INDUSTRIAL INFRASTRUCTURE

1.1 Introduction

The Industrial Internet of Things (IIoT) expresses itself as a subset of Internet of Things (IoT) technology within the usage of the industrial sectors and applications. In recent years, especially after the COVID-19 pandemic, information and communication technology (ICT) has changed the way industrial control systems (ICS) operate precisely in their networks. Such a change has brought the implementation or application of new communication standards in the direction of openness, including computer and wireless networks. Over the years, this change has continued, and it has raised many questions about the security and cyber security aspects of these industrial infrastructures. Moreover, different cyber threats in form and content have appeared and, in this way, have complicated industrial protection from these cyber threats. These cyber threats are constantly being verified by many cyber-attacks, including ransomware, malware, phishing, and distributed denial of service DDoS attacks, which are successfully managed to bypass the security of ICS systems. There has been demand from the industry and the scientific community in recent years; a commitment has been observed for a security solution that would guide the needs of the industry for advancing the level of cyber security in their systems and networks.

Many issues require solutions regarding cyber security for ICS and emergent technologies, which present good opportunities but also bring new risks. The purpose of this chapter is to show the main purpose of highlighting the challenges in the field of cyber security to secure ICS. Cloud infrastructures have brought many innovations to the market in terms of different tools, methods, and software that enable a more advanced operation of the industry and its infrastructures. Cloud technology greatly supports ICS and supervisory control and data acquisition (SCADA) systems through the construction of different environments according to public or private needs. Based on the general principles of IIoT security, it is worth mentioning the following: secure IIoT systems before integration, network segmentation, monitoring and analysis, attribution of cyber-attacks, detection tools for malicious activities, and system maintenance regularly. Figure 1.1 illustrates an example of IIoT network separation in the layer.

The big change that is happening, or better said, the digitization of objects in smart ones as devices, enables a historical advancement and shows an approach to human interaction with the Internet.

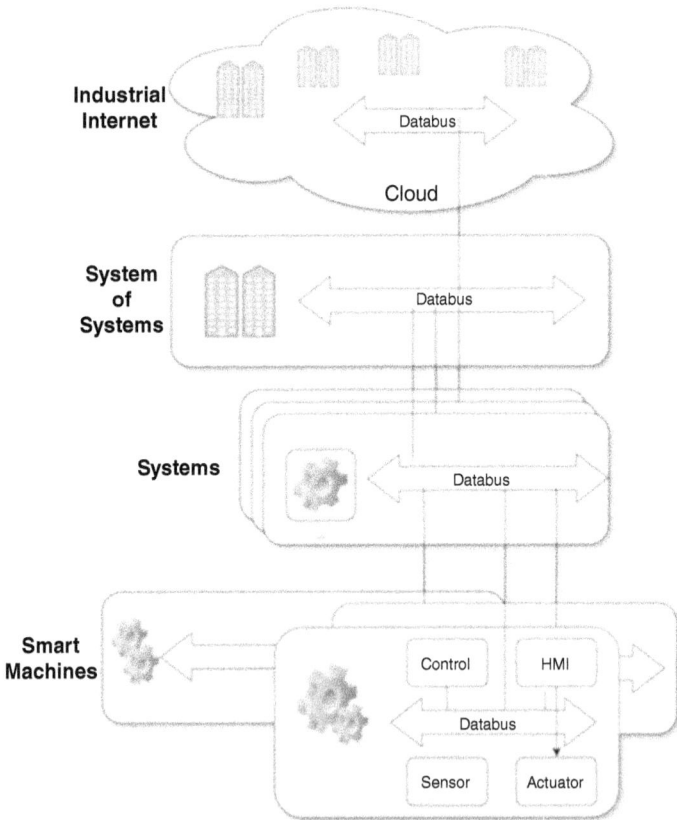

Figure 1.1 Example of IIoT network separation.

The book's focus, are IoT sensors and their industrial subset, the IIoT, which enable the communication of the physical part with the virtual part, which in emerging technologies is known as the digital twin. The IIoT includes the use of the IoT in industry and applications; since the impact of big data, machine learning, and machine communication is great, IIoT helps the industry in the efficiency of operations and sustainability. Figure 1.2 presents an example of the architecture of IIoT.

The application of the IoT and the IIoT are information systems and devices that are bringing a new business model into use, where IoT systems rely heavily on the Internet, i.e., IP communication protocols.

The rest of the chapter is structured presentingthe definition section explains the meaning of information technology (IT), industrial sensors, and information systems and presents an overview of cyber-security within the industry concerning threats and cyber-attacks. In the following section, the timeline of the IoT presents its evolution toward a new subset like the IIoT.

1.2 Definition: Cyber-Physical Systems

The cyber-physical system (CPS) is an integration of computation of physical processes where it's a combination of behaviors defined in the cyber and physical world. The term "cyber-physical systems" was first used by Helen Gill in 2006 in USA. The term CPS is often mentioned and associated with the IIoT, Industry 4.0. A CPS represents a system with computing and networking components. Different examples can present a CPS structure, but Figure 1.3 presents one divided into parts.

The first part is the physical plant, and the CPS can include mechanical or biological parts. The second part is one or more computational platforms consisting of sensors. The third network fabric provides a mechanism for communication. Finally, the platform and network fabric create the cyber part of a CPS.

1.3 Definition: Information Technology (IT)

Since its beginning, technology has continuously transformed our lives, organizations, industries, and countries. The human being is its creator and witness to the great events that technology has brought

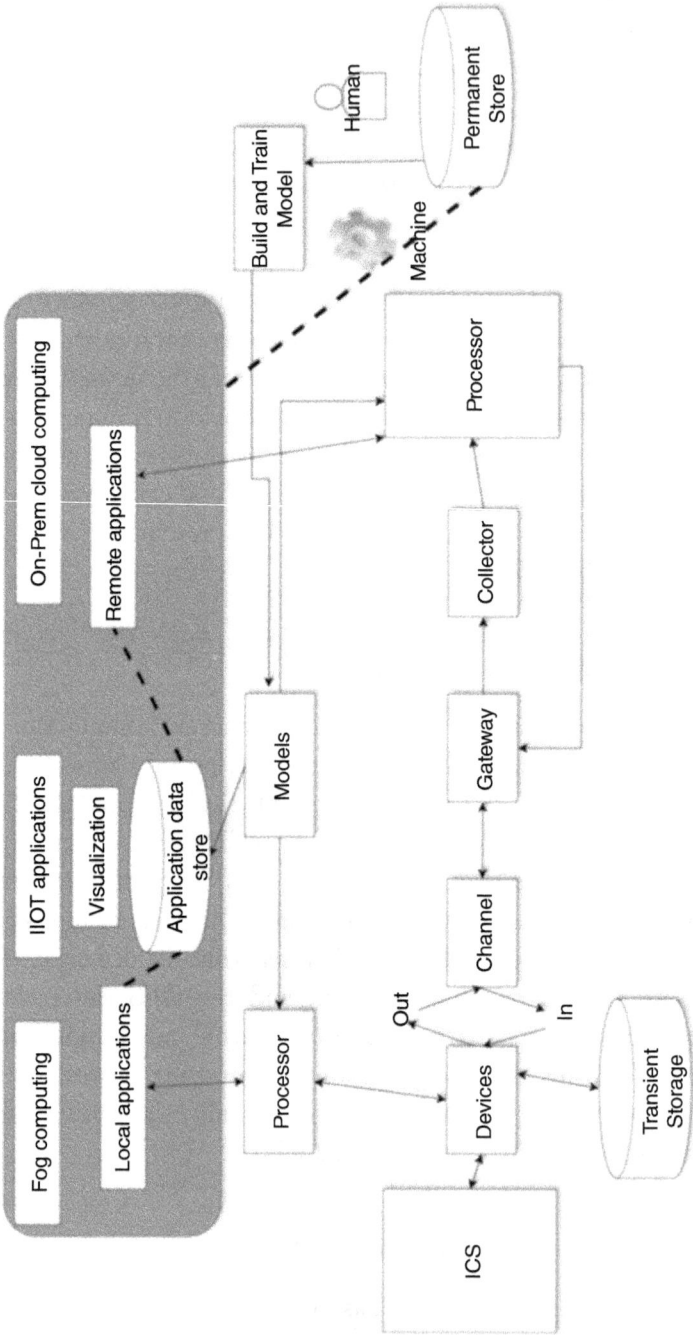

Figure 1.2   The IIoT architecture.

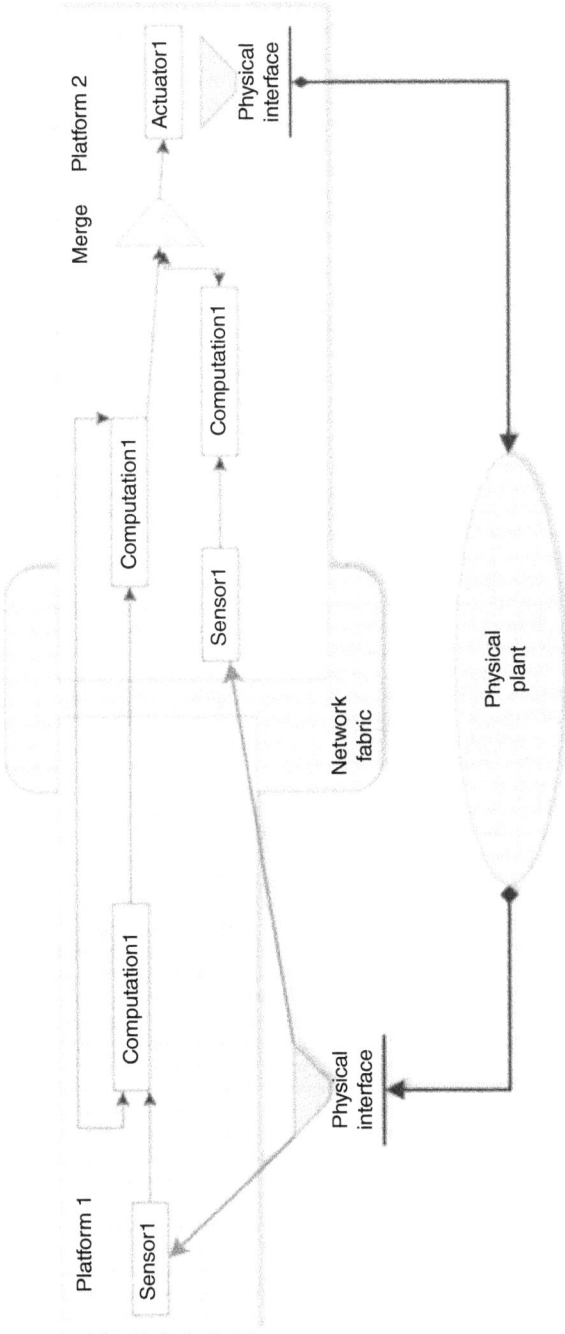

Figure 1.3 Example of cyber-physical system structure.

with its application in many sectors of our lives. Information technology (IT) refers to forms of technology in the processing, storage, and transmission of information in electronic or digital form. For this reason, computer equipment such as computers, other communication devices, and networks is used.

1.4 Definition: IoT

The IoT is the network of physical sensors that contain embedded technology to communicate and sense or interact with their internal states or the external environment of the Internet. The construction of IoT devices and their connection in a large network of millions of such devices creates a smart world where physical and digital parts become one. If these devices are found in the Internet world and their interoperability is possible with people, computers, and other intelligent parts, it also represents a cyber risk or threat. The possibility of compromising IoT today is very large because their manufacturers do not care at all that about the protective measures implemented in hardware or such instructions. As shown in Figure 1.4, IoT is divided into several main layers, including a device layer, a network layer, a support layer, and the application layer. IoT devices are presented in different sensor categories, including physical, chemical, and biological sensor. These

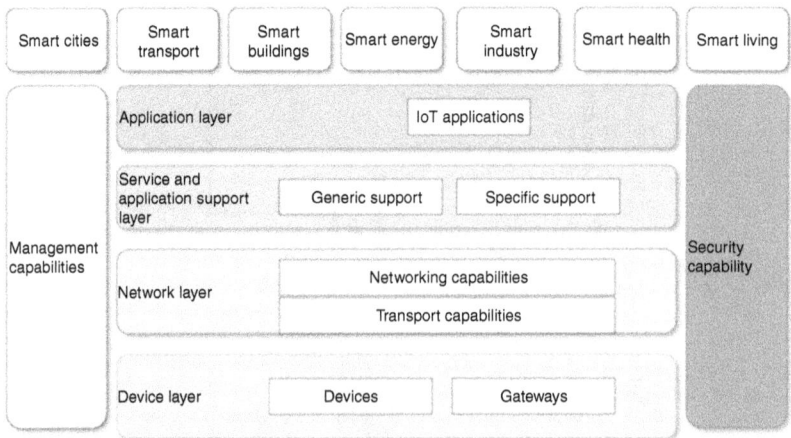

Figure 1.4   IoT layers architecture.

sensors can also absorb information from the physical world for certain purposes, processing and sending the data that have been built.

1.5 Definition: Industrial Internet of Things (IIoT)

The IIoT refers to the use of sensory devices to enhance industrial processes across various sectors. The development of the IoT has enabled the connection of numerous devices in both business and industry. The term IoT encompasses the use of digital technology in industry, which has led to the adoption of the term industrial IoT, which is considered a subset of IoT devices. IIoT is utilized in industrial sectors and emphasizes machine-to-machine (M2M) communications, machine learning, and big data. Recently, IIoT has been divided into three categories: industrial IoT, commercial IoT, and consumer IoT, with further divisions continuing to emerge. This book focuses on industrial IoT, particularly concerning cybersecurity issues.

1.6 Definition: Industrial Information Systems (IIS)

The information system (IS) consists of integrated components that collect, store, and process data. Various types of IS are utilized in industry, including knowledge systems, management systems, and decision support systems. There are numerous applications of industrial information systems (IIS); currently, several industrial processes are integrated within these systems, such as customer relationship management (CRM), order management, warehouse management, and human resources.

1.7 Different Types of Cybersecurity

Cyber security can be divided into several main categories, including critical infrastructure, application security, network security, cloud security, and IoT security. All this connection of the world through different devices on the Internet and the creation of a giant communication network also comes with risks and cyber threats. Knowing this, cybersecurity has received focus and present an issue that is discussed and always needed against cyber threats, by using appropriate measures for the protection. Regardless of what countermeasures are put in place, a person is the most vulnerable point, and if he does not follow the rules, there will always be a security breach.

To cover all these main divisions of cybersecurity, three components, including people, processes, and technology, must be addressed. All frameworks, standards, and methodologies of cybersecurity and security foresee how to handle these three components through different methods of implementing security policies.

Improving software security is always a continuous process for software vendors. If you have software that is used by many users, every security weakness represents a risk for the product and its customers. Even with the right security updates, you can never be 100 percent patched.

Cybersecurity is not limited and is treated only as a technical problem; it must go further and be approached from different aspects. In cybersecurity, building and cultivating the organizational culture around cybersecurity and people's awareness is important. Understanding the rationality of the cybersecurity context involves moving away from the myth of cybersecurity, cyber synonyms, the combination of people and machines, and security, the freedom from risk or danger. From space, security weaknesses in infrastructure, equipment, or systems bring the success of cyber-attacks, which are usually carried out by individuals or organized groups known as advanced persistent threats (APTs).

1.8 Timeline of IoT

The Fourth Industrial Revolution has made the great integration between IoT devices and the Internet, creating a wide communication network. Smart IoT devices have advanced and can collect different data for positions, locations, patterns, etc. IoT devices are constantly being advanced; they have the ability not only to collect information but also to share it with others. The concept, logic, and architecture of IoT have been very successful, further developing other subsets such as IIoT, which is the main pillar of Industry 4.0 in our time.

A chronology of the developments of the industrial revolutions, from the agrarian one to the technological one, is presented as follows. Industrial revolutions started with mechanization through steam and water. The development of the second revolution was the advancement from steam power to electricity. Advances in electronic components and technological circuits produced the computing devices that enabled the third industrial revolution. Known as Industry 4.0, the Fourth Industrial Revolution is presented through the interconnection of devices and their advancement toward intelligence, and the development of artificial intelligence characterizes this revolution.

1.9 IoT to IIoT Issues, Challenges, and Benefits

The digital transformation every day and more from the events happening in cyberspace toward critical information infrastructures, industry, and its infrastructure, including the IoT, requires a higher level of cybersecurity. Issues and challenges have evolved and advanced since 2020. Still, some remain the same, including insufficient prioritization of cybersecurity in security programs and a lack of a clear definition of security among senior executives. Cybersecurity benefits remain, in all this rapid development of emergent technologies and the Fourth Industrial Revolution, cybersecurity it is bringing many opportunities to advance our concepts, approaches, and work methods.

As mentioned, the developments of IoT and its subset, industrial IoT, Industry 4.0, have brought a new approach, a concept known as digital twins. A digital twin is a presentation and representation of a physical object in the virtual world with all the attributes of the physical one. The digital twin will completely change the concept of the industry's operation in the coming years.

The industry and its sectors, including production, transport, sales, and health, have recently benefited from and applied technological connections for IoT and industrial IoT devices. The main challenge of the time remains securing IIoT in industrial infrastructures from cyber threats. The consequences of cyber-attacks on IIoT and industry are having an impact on compromising the IIoT infrastructures and its normal operations.

1.10 Industrial Internet of Things (IIoT)

The IIoT devices enable their interconnection with the Internet, collecting and transmitting data of the physical part that they represent.

The IIoT is an extension of the use of the IoT in the industrial and application sectors, focusing on M2M communication. The application of IIoT in industry, including health, sales, logistics, vehicles, and oil refineries, has facilitated their operation in terms of communication between devices and providing information on the spot for decision-making. The recognition or application of IoT in almost all industry sectors has brought good things but also challenges that are still being identified over time. The issues that have opened a discussion that preoccupies the industry are security weaknesses, connectivity, legal and regulatory issues, and lack of a standardized architecture. The future of IIoT is very promising but also presents unidentified risks; its application will expand and advance toward integration with artificial intelligence, the digital twin concept, etc.

The IIoT network created by the many connections that they have among themselves consists of physical devices, systems, applications, etc. All this has taken off in the time of Industry 4.0, which is known as the time of digital transformation of everything.

In terms of the security of IoT devices, the analysis can be classified at the level of information, access, and functional. The layers of IoT are classified into edge, access, and application, which supports the focus on weaknesses.

1.11 Lifecycle of an IIoT

The IIoT has reached its development for a short time, but its use has taken off in the industry. It has managed to do different things in the industry by increasing productivity and effectiveness. With the lack of secure network encryption, IIoT implementation can bring security challenges and vulnerabilities. Cyber security is vital for the implementation and normal operation of IIoT. So, the IoT is getting a lot of attention, and its definition and description by The Internet Engineering Task Force (IETF), says this *"in the vision of the IoT, 'things' are very various such as computers, sensors, people, actuators, TVs, etc."*.

During the lifetime of IIoT, threats and threat actors have opportunities to exploit IIoT and compromise their security. Threats to IIoT systems can be cyber-attacks, botnets, APT organizations or individuals, and insiders.

But some of the ongoing and recent cyber-attacks remain these:

- Ransomware
- Malware
- Denial of service attacks
- Access control attacks
- Phishing emails

Two things are taken into consideration before launching a cyber-attack on any target: attack surface and attack vector. These two are related to the industry, the attack surface is related to the system components. The attack vector includes the use of methods, tools, and technologies to carry out the cyber-attack. For dealing with cyber-attacks, including forms, methods, tools, technologies, and targets, different methodologies are used that facilitate the handling of incidents through their instructions. The Open Web Application Security Project OWASP methodology is well-known, especially for IoT attack surface models, and it lists identified cyber-attacks and the vulnerabilities that cause them. In addition, we will mention other methodologies and models in the next chapter and talk about them in detail in terms of handling security vulnerabilities.

Suggested Websites

- National Institute of Standards and Technology (NIST) https://www.nist.gov/
- Cisco Networking – Industrial Networks https://www.cisco.com/
- Gartner IT Glossary https://www.gartner.com/en/information-technology

2

ARCHITECTURE AND COMMUNICATION IN AN INDUSTRIAL INFRASTRUCTURE

2.1 Network Industrial Architecture

In the industrial infrastructure, the operation and support of applications are challenging. There is a need to constantly enable communication between components, applications, devices, and data exchange from different sources; for this, the operation of the network is important. Industrial communication protocols for Industry 4.0 require continuous communication with different factory components; communication protocols that are now considered old were used in the past (Figure 2.1).

Furthermore, network architecture represents the way services and devices are grouped to serve and interconnect the needs of different clients and applications. One of the most important aspects of network architecture, among others, is security. Undisputed, but with a significant impact on the architecture of the network in its operation, security is one of the foundations. Security is addressed in all aspects of the network infrastructure, from physical security to data transmission protection, on-premises storage, and Cloud environments.

Industrial networks are essential in the control process; they go beyond the concept of interconnection between factory systems. Furthermore, the selection or definition of industrial networks is based on several conditions that determine requirements, capacities, accessibility, costs, and maintenance. In industrial networks, the lack of standardization represents a serious problem in communication.

Figure 2.2 shows the different parts of the industrial network architecture within the industrial environment. Each part is important for functioning as a whole network architecture, from the Cloud

 DOI: 10.1201/9781003383253-2

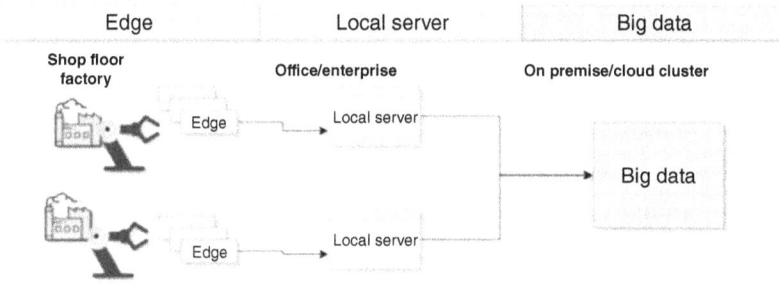

Figure 2.1 Architecture presentation.

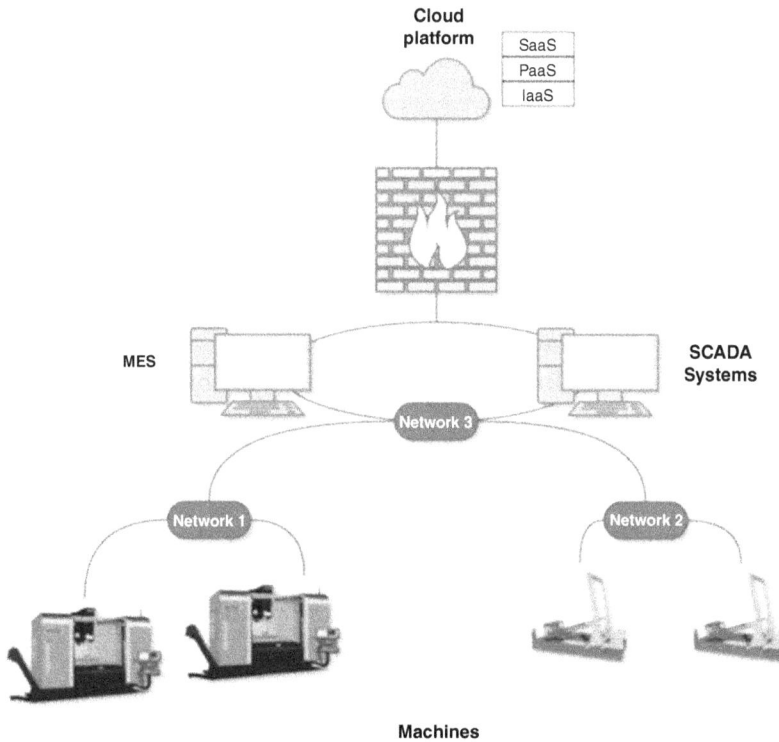

Figure 2.2 Industrial networking architecture introduction.

environment, firewall protection, and industrial systems, to the networks where machines are used to communicate and transmit data.

The entire architecture of the Industrial Internet of Things (IIoT) system requires the operation, access, readiness, and stability of industrial networks in such a way that the data is transmitted while respecting and protecting their integrity. New technologies are bringing innovations to industrial networks as well. The 5G technology that has just been put into operation and the 6G technology will make a big difference in computer networks, and the operation of the industry will advance further toward a unified network. For future distributed communications and processing architecture, market segments will benefit from 5G technology.

World industrial economies are experiencing the process called the Fourth Industrial Revolution, Industry 4.0, which includes several developments: digitization of factories, use of big data, Internet of Things (IoT), smart cities, etc.

2.1.1 Models of Communication

When it comes to communication models, emerging technologies are bringing new ideas of communication applications for IIoT. The evolution of IIoT and Industry 4.0 has brought opportunities to advance communications between robotic tools and their systems; regardless of these, there are concerns about their security. Driven by new technologies, developments in communications have advanced and will continue to change in the future. The developments in the connection of the industry in every aspect of the transformation toward the advancement of the factories in the future have a great impact on the industrial automation infrastructure.

Communication models in the industry involve the transmission of data from the work environment to the system control level. Proper and good communication enables understanding between workers and machines through systems where they all have a common point. For the security to be at the right level and the infrastructure and the organization to be safe, it is recommended that the business network be separated from the industrial network where sensors, machines, etc. are connected. This separation or isolation of organization and industry networks is aimed at preventing cyber-attacks that can pass or penetrate the networks. Industry networks can be CAN (Controller Area Network), SCADA (Supervisory Control and Data Acquisition), Ethernet, and LAN (Local Area Network), normally OSI (Open Systems Interconnection) models.

2.1.2 Architecture of the Industrial Internet of Things

The IIoT, a subset of Internet of Things (IoT), has reached its development in the era of Industry 4.0. The architecture of IIoT can be presented like this in Figure 2.3, divided into layers. There is no single standard or way to deploy Internet of Things (IoT) infrastructure, but it considers the users' perceptions and the applications that will be used. Figure 2.3 includes the physical component, communication, i.e., devices, data storage, and the interactive layer at the fourth level. Concerning the IIoT, the Industrial Internet Consortium (IIC) organization has been working on a framework for the IIoT, known as the Industrial Internet Reference Architecture (IIRA).

The Industrial Internet Reference Architecture (IIRA), as a common IoT framework, helps in the interoperability of IIoT systems for different applications. Its purpose is to support and assist the private and public sectors in the development, documentation, communication, and implementation of IIoT systems. The challenges, problems, and issues that are worrying the whole community and industry in IIoT are different and several. Before starting the implementation of IIoT in other components, it is important to address the challenges that include security, preserving the visibility of "Things", industrial

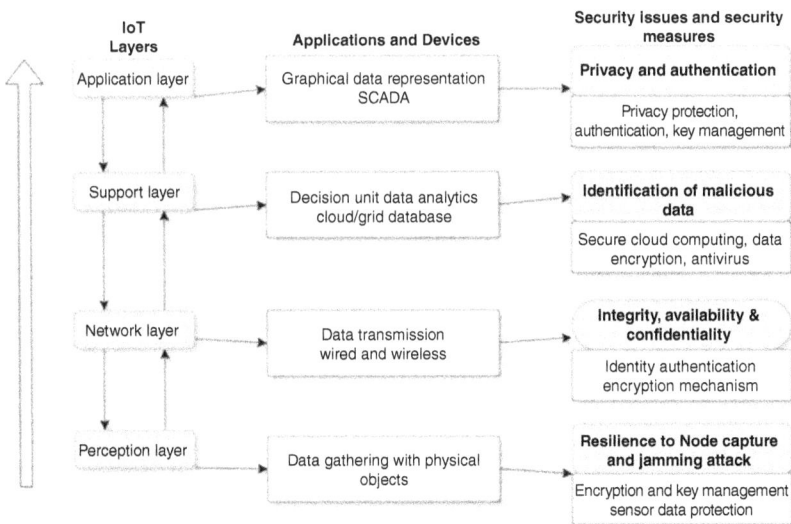

Figure 2.3   IIoT architecture layers, applications, security.

Ethernet protection, unified legacy systems and IIoT Infrastructure, and data storage.

2.2 Communication Network

IoT communication includes the infrastructure, technology, protocols, and gateways that enable successful data transmission. The access network of the industrial Internet is a transport network that connects networks and protocols.

2.3 Transport Network

The continuous change in the use of Ethernet services and the constant rise in demand for smart devices have also affected the transport network. The main parties in the optical transport network market are Cisco, Fujitsu, Nokia, etc. Industry 4.0 represents a big change in the industry, almost all its sectors. New technologies such as 5G and 6G will have a key role in supporting low-cost transportation compared to the technologies used so far. Experiments for the application in the industry are being done with optical networks to advance data transmission somewhat like in the air.

Networks, which consist of links and nodes, are the basis for the function of IIoT infrastructure, applications, and sensors.

2.4 Protocols on the Internet

Protocols are the foundation of Internet communication, facilitating data exchange between devices, servers, and networks. In this section, we explore key internet protocols that enable the seamless transmission of information across the digital landscape.

2.4.1 Transmission Control Protocol (TCP)/Internet Protocol (IP)

TCP/IP is the foundational protocol suite of the Internet, providing reliable and connection-oriented communication between devices. TCP ensures that data packets are delivered in sequence and without errors, while IP handles the routing of packets across networks. Together, TCP/IP forms the backbone of Internet communication, enabling devices to communicate across disparate networks.

2.4.2 *Hypertext Transfer Protocol (HTTP)/Hypertext Transfer Protocol Secure (HTTPS)*

HTTP is the protocol used for transmitting web pages and other resources over the Internet. It operates as a request-response protocol, where a client sends a request to a server for a resource, and the server responds with the requested data. HTTPS is a secure version of HTTP that encrypts data transmission, providing confidentiality and integrity for web communications, crucial for secure transactions and data exchange.

2.4.3 *File Transfer Protocol (FTP)/Secure File Transfer Protocol (SFTP)*

FTP is a protocol used for transferring files between a client and a server on a network. It enables users to upload, download, and manage files remotely. SFTP, an extension of SSH (Secure Shell), provides secure file transfer capabilities by encrypting data during transmission and protecting sensitive information from interception or tampering.

2.4.4 *Simple Mail Transfer Protocol (SMTP)/Post Office Protocol (POP)/Internet Message Access Protocol (IMAP)*

SMTP is the standard protocol for sending email messages between servers. It defines how email messages should be formatted, transmitted, and delivered to the recipient's email server. POP and IMAP are protocols used by email clients to retrieve messages from a server. While POP downloads messages to the client device, IMAP synchronizes messages across multiple devices, ensuring consistent access to emails from anywhere.

2.4.5 *Domain Name System (DNS)*

DNS is a distributed naming system that translates domain names (e.g., example.com) into IP addresses, allowing users to access websites and services using human-readable names. DNS plays a crucial role in Internet navigation, enabling users to locate resources and servers without needing to memorize complex numerical addresses.

2.4.6 *Secure Shell (SSH)*

SSH is a cryptographic network protocol used for secure remote access to devices over an unsecured network. It provides encrypted communication between a client and a server, preventing eavesdropping and unauthorized access. SSH is commonly used for remote administration, file transfer, and tunneling, offering a secure alternative to traditional remote access methods.

Table 2.1 shows how various internet protocols are utilized in industrial settings, including IoT and IIoT applications, demonstrating their critical roles in enabling communication, security, and data exchange within industrial environments.

Table 2.1   List of Various Internet Protocols Along with Their Applications in Industry, Including IoT and IIoT

PROTOCOL	DESCRIPTION	INDUSTRY APPLICATIONS
Transmission control protocol (TCP)	Provides reliable, connection-oriented communication between devices.	IoT data transmission; IIoT device communication.
Internet protocol (IP)	Handles the routing of packets across networks.	IIoT network routing; IoT device addressing.
Hypertext transfer protocol (HTTP)	Transmits web pages and resources over the Internet.	IIoT web-based monitoring interfaces; IoT device management.
Hypertext transfer protocol secure (HTTPS)	Secure version of HTTP, encrypts data transmission.	Secure data exchange in IIoT applications; IoT device communication over encrypted channels.
File transfer protocol (FTP)	Transfers files between a client and a server.	IIoT data backup and transfer; firmware updates for IoT devices.
Secure file transfer protocol (SFTP)	Secure file transfer protocol, encrypts data transmission.	Secure data transfer for IIoT devices; encrypted firmware updates for IoT devices.
Simple mail transfer protocol (SMTP)	Sends email messages between servers.	IIoT device notifications; email alerts for IoT systems.
Post office protocol (POP)	Retrieves email messages from a server, downloading to the client device.	Email retrieval for IIoT devices, POP-based email notification systems for IoT applications.
Internet message access protocol (IMAP)	Synchronizes email messages across multiple devices.	IMAP-based email synchronization for IIoT devices; email access from IoT devices.
Domain name system (DNS)	Translates domain names into IP addresses.	IIoT device addressing, DNS resolution for IoT device communication.
Secure shell (SSH)	Provides secure remote access to devices over unsecured networks.	Secure remote management of IIoT devices, and encrypted communication for IoT device administration.

Protocols form the backbone of Internet communication, enabling the seamless exchange of data and services across diverse networks and devices. Understanding these protocols is essential for navigating the complexities of the Internet and ensuring secure and efficient communication in an interconnected world. As technology evolves, new protocols may emerge, further shaping the landscape of Internet communication and connectivity.

2.5 Industrial Protocols

Communication protocols in the industry are an important aspect and will be addressed in this chapter in more detail by introducing them. Communication is now only used in distributed computer systems. In continuation of the development of Ethernet networks, also wireless networks, i.e., wireless technology is being widely used in the industry. Communication systems in the industry can be categorized as follows: real-time behavior, distribution, homogeneity, and installation type. Wireless networks in the industry have achieved good use and are known as wireless local area networks (WLANs). In industrial automation processes, there is a wide discussion and research regarding wireless sensor networks (WSNs).

Although emerging technologies have had a great impact and raised the issue of finding a solution for industrial protocols and networks, industrial automation continues to be a trend.

2.6 Industrial Internet of Things (IIoT) Protocols

The Industrial Internet of Things (IIoT) represents a big change in the industry, especially in the production sector. IIoT devices and sensors are in high demand and are used by the industry. IIoT, known in Industry 4.0, has more precisely transformed the classic factories as we know them into smart ones. Which now include in their infrastructures machine learning, big data, and machine-to-machine (M2M) communications. While IoT is reaching the peaks of use in the health sectors and smart cities, IIoT is reaching the scope of use in industries and their infrastructures. Most of the industry sectors that have applied IIoT include manufacturing, transportation, energy, health, oil, refinery, etc.

Feature	HTTP	CoAP	MQTT	MODBUS TCP
infrastructure	Ethernet, Wi-Fi	6LoWPAN	Ethernet, Wi-Fi	Ethernet, Wi-Fi
network layer	IPv4 or IPv6	IPv6	IPv4 or IPv6	IPv4 or IPv6
transport layer	TCP	UDP	TCP	TCP
transport port	80, 443	5683	1883, 8883	502, 802
model	synchronous	asynchronous	asynchronous	synchronous
pattern	request—response	both	publish—subscribe	request—response
mechanism	one-to-one	one-to-one	one-to-many	one-to-one
methodology	document-oriented	document-oriented	message-oriented	byte-oriented
paradigm	long polling-based	polling-based	event-based	polling-based
quality level	one level	two: CON or NON	three: QoS 0,1,2	one level
standard	IETF (RFC7230)	IETF (RFC7252)	ISO/IEC, OASIS	modbus.org
encoding	ASCII text	RESTful (Binary)	UTF-8 (Binary)	Binary
security	SSL, TLS	DTLS	SSL, TLS	TLS

Figure 2.4 Comparison of IoT protocols.

The development of Industry 4.0 or IIoT aims to extend technology to all different industrial processes and machines for automation and operational efficiency.

The IIoT requires communication with devices and information systems to transmit the collected information to them. The main application protocols in use for IIoT are Hypertext Transfer Protocol (HTTP), Message Queuing Telemetry Transport (MQTT), Advanced Message Queuing Protocol (AMQP), Constrained Application Protocol (CoAP), etc.

Figure 2.4 introduces some comparisons between some of the IoT protocols in the aspect of infrastructure, architecture, mechanism, model, messaging pattern, methodology, and transmission.

The MQTT (message queuing telemetry transport) protocol is a messaging protocol and is often used in IoT sensor devices. The MQTT protocol is suitable for machine-to-machine communications.

2.7 Other Protocols

In the list of IoT communication protocols, other protocols are used in different cases. These include data distribution service (DDS), WebSocket, advanced message queue protocol (AMQP), extensible messaging and presence protocol (XMPP), and OPC unified architecture (OPC UA). Communications of the protocols used by IoT can be done through satellites, WiFi, radio frequencies, Radio-Frequency Identification (RFID), Bluetooth, and Near-field communication (NFC). Internet protocol (IP) is a set of rules that dictates how data gets sent to the Internet. IoT protocols ensure that information from one device or sensor gets read and understood by another device, a gateway, or a service.

Suggested Websites

- Industrial Internet Consortium (IIC) https://www.iiconsortium.org/
- 4th Industrial Revolution – Industry 4.0 https://www.weforum.org/focus/fourth-industrial-revolution/ OPC Foundation – OPC Unified Architecture (OPC UA) https://opcfoundation.org

3

TRADITIONAL SECURITY

3.1 Introduction

Traditional security is a technological practice used to protect information systems through countermeasures. These countermeasures are designed to serve as protection from cyber-attacks and threats from different directions that may come, including hackers, viruses, and internal and external threats. The initial and main part of the security that protects us from the outside world on the Internet is the network protection devices firewalls. The purpose of firewalls is to filter all incoming and outgoing traffic based on a set of security rules. However, even for threats from malicious codes and viruses, different software known as antiviruses are used, the main functions of which are the detection and deletion of these threats from the system. There is also an advancement in this software also presented as intrusion detection and prevention systems, which are used by organizations as a solution to block unauthorized access to the network. Traditional security also includes other ways of countermeasures, which will be described in more detail in the following chapter. In addition to technical issues, the non-technical issues that traditional security foresees are also important, including security awareness programs, policies, and incident management plans. These countermeasures serve sufficiently to raise the awareness of the staff in the organization for security and their response in the event of an incident or potential threat. In general, traditional security represents the leading component in organizations, that is, IT infrastructure, where it has its roots, and through technical protective countermeasures, they regularly audit and maintain security.

DOI: 10.1201/9781003383253-3

3.2 Objectives

The fundamental purpose of security in information technology (IT) is to protect us from misuse, theft of information, unauthorized access, and or damage through the involvement of technical and non-technical countermeasures. These protections include facing external and internal threats, e.g., hackers, viruses, unauthorized access, and misuse of resources. The central and other objectives of traditional IT security are:

- Providing confidentiality, integrity, and availability for information and systems, known as the CIA triangle.
- Compliance with legal aspects, regulations, and industrial standards.
- Protection from security breaches, security incidents that can cause damage to the organization's image, and financial loss.
- Security awareness programs provide the opportunity to understand security and raise awareness among all the organization's staff. This helps a lot in avoiding security threats or incidents.

To achieve the level of security, its maintenance as the main objective of traditional security in IT is supported by the combination of technical countermeasures such as firewalls, antivirus, access control, and non-technical security awareness programs. In general, traditional security in IT has to do with protection from threats and attacks that target the confidentiality, integrity, and availability of systems and information.

3.3 Levels of IT Security

There are different levels and some of IT security that can be used to protect against threats and attacks on systems and information. From all those levels of IT security, we distinguish the following ones depending on the physical, network, endpoint, data, and application aspects.

- Physical security refers to the physical protective measures that are put in place to prevent physical access to computer systems and data. These measures can include server rooms with locks, cameras, and access control.

- Network security includes protection against unauthorized access or network attacks. These protective measures can be provided using firewalls, detection and prevention systems known as IDS/IPS, and virtual private networks (VPNs).
- Endpoint security includes the protection of devices that are connected to networks, such as PCs and servers. Here, we can use protective measures such as antivirus, device and application control, and mobile device management (MDM) systems.
- Data security is crucial for the continued operation of information systems; protective measures include protection against unauthorized access or change. Protective measures against this are encryption, access control, backup copies, and system recovery.
- Application security includes many things, but we have some countermeasures to protect against software and application vulnerabilities. Countermeasures include code review, input parameter validation, and code security best practices.

In general, it plays a very important role in IT security for organizations to implement a comprehensive approach to the security program that addresses all levels of IT security and offers solutions for them, with an unchanged focus on providing security and protection from threats and attacks that target information and systems (Table 3.1).

If we further address the security level of the application and we give a clearer understanding with examples of measures that can be put in place to protect against software and application vulnerabilities. Measures that can be included in this level of application security include code reviews, input validation, and secure coding practices. The application is important in IT security and for the organization because the way it is built depends much more on the prevention of software and application attacks that can compromise data and the system. Some examples that are currently used in the industry in application security measures include:

- Code review is the process where the developer reviews and tests the code to find and fix security vulnerabilities before the product is released to production. Even according to different standards and approaches to development methodologies,

Table 3.1   Outlines the Different Levels of IT Security Across Various Aspects

SECURITY ASPECT	SECURITY LEVEL	DESCRIPTION
Physical security	Level 1	Basic physical security measures, such as locks on doors and windows, are in place.
	Level 2	Security cameras, access control systems, and other physical security measures are implemented.
	Level 3	Advanced physical security measures, such as biometric authentication and intrusion detection systems, are used.
Network security	Level 1	Basic network security measures, such as firewalls and antivirus software, are in place.
	Level 2	Intrusion detection and prevention systems, virtual private networks (VPNs), and other advanced security measures are implemented.
	Level 3	Advanced network security measures, such as next-generation firewalls and security information and event management (SIEM) systems, are used.
Endpoint security	Level 1	Basic endpoint security measures, such as antivirus software, are in place.
	Level 2	Endpoint detection and response (EDR) systems, whitelisting, and other advanced security measures are implemented.
	Level 3	Advanced endpoint security measures, such as behavioral analysis and artificial intelligence/machine learning-based systems, are used.
Data security	Level 1	Basic data security measures, such as access controls and encryption, are in place.
	Level 2	Data loss prevention (DLP) systems, backup and recovery procedures, and other advanced security measures are implemented.
	Level 3	Advanced data security measures, such as data classification and tagging, data encryption at rest and in transit, and secure key management, are used.
Application security	Level 1	Basic application security measures, such as password policies and access controls, are in place.
	Level 2	Web application firewalls (WAFs), vulnerability scanning, and other advanced security measures are implemented.
	Level 3	Advanced application security measures, such as penetration testing and code reviews, are used.

such a process is foreseen, but often organizations do not define them in detail and inadvertently cause security vulnerabilities in their software and applications.

- Input validation verifies and checks the data that is set as input by the simple user, and a comparison is made to see if it meets the defined criteria. Usually, this method or technique prevents malicious code injection, e.g., SQL injection attacks that aim and target the database and the acquisition of data or their manipulation.

- Secure coding practices follow the best practices of writing code that increase the level of security. Different functions, validations, and error handling can be used here.

Furthermore, the implementation of strong application security measures is required to protect against software and application vulnerabilities. To maintain the security level of confidentiality, integrity, and availability of data and systems, application security measures are more necessary in the time we live in and the trend of technological developments in the industry. This is also due to the increase in support for software products and applications of organizations and the increase in threats and cyber-attacks. In the operations of an organization, software and applications have a key role due to their usability and reliability. At the same time, they often become targets for attackers with the sole purpose of gaining access and disrupting the normality of operations. Without clear and strict policies and rules of application security measures within the organization, the risk is high from the big losses that can come from threats and cyber-attacks. Putting in place or implementing stringent application security measures, policies, and rules helps protect against vulnerabilities that may be in software and applications. Some of these application security measures are mentioned earlier in the chapter, including code reviews, input validation, and secure coding practices, which prevent attackers from exploiting the vulnerability.

In general, it is fundamental that the application of security measures in the industry cannot be avoided, and high priority should be given to organizations in the strict implementation of policies and rules of security measures in defense against growing threats and cyber-attacks.

3.4 Principle

The basic principles of IT security are the foundation of the practical concept of protecting data and systems from threats and risks. The basic principles include as follows:

- Confidentiality – Confidentiality involves protecting information that is not disclosed to the public by unauthorized access or parties. The utilization of measures such as encryption and access control are ways to achieve confidentiality.
- Integrity – Integrity protects data and systems from tampering or unauthorized changes. Utilization of measures checksums, hashing algorithms, and access controls enable integrity.
- Availability – Availability is presented as the possibility of authorized access and achievement over the network by users to data and systems. Utilization of measures such as redundant systems, disaster recovery plans, and load balancing provides availability.
- Authenticity – Authenticity is done by verifying and identifying the user who he really is. It can be achieved by the utilization of measures such as passwords and two-factor authentication.
- Non-repudiation – Non-repudiation has to do with the process of proving which action was performed and by whom, that is, the user or individual; this can be achieved through digital signature measures (Table 3.2).

As emphasized at the beginning of this paragraph, the principles of IT security are essential for the protection of data and systems from threats and attacks. In addition to the technical principles of

Table 3.2 Common IT Security Principles

IT SECURITY PRINCIPLE	MEANING
Confidentiality	Ensuring that only authorized individuals or systems have access to sensitive information.
Integrity	Ensuring that data is accurate, complete, and trustworthy.
Availability	Ensuring that authorized users have access to information and systems when they need it.
Authenticity	Ensuring that data or information is genuine, and its origin or source can be verified.
Non-repudiation	Ensuring that a user cannot deny having performed a specific action or accessed specific data.

IT security mentioned above, we also have non-technical principles that are important in the security of information and systems. These include:

- Risk management is a process of identifying, evaluating, and managing risk through the implementation of controls, continuous review, and update of risk assessment.
- Policies and procedures define rules and guidelines for the industry that must be followed to have a level of security for information and systems.
- Employee training is a user education program to be aware of their actions in terms of data and systems security. These are usually realized through security awareness programs, which educate the organization's staff on the topics of password management and recognizing phishing attacks.
- Incident response builds plans and strategies that show how to act in cases where the incident occurs or has occurred. These plans clearly define the actions and activities that must be undertaken to minimize the loss of the organization.

These non-technical principles provide a level of security in other aspects that complement the technical one and regulate important aspects of the organization and industry in defense against threats and attacks.

3.5 Security Governance on Industrial Infrastructure

Security governance deals with the rules, processes, and policies that are put in place to provide and enhance the security of the industrial infrastructure. Who are the industrial infrastructures and what is included in them? They include the systems and networks used for monitoring and control of critical infrastructures, which can be water distribution systems, production plants, energy, etc. Everything related to these and around them is considered critical infrastructure, and in case of compromise of one of them, they have a great national impact on a country. To make sure that the level of security is maintained and where it is required, we must have security governance that prevents major losses that may be a consequence of threats and attacks. Threats that target these critical infrastructures can also cause loss of life.

Security governance in organizations should be done through the implementation of some solutions, such as the technical and non-technical aspects of security. These solutions include protective measures such as firewalls, intrusion detection, and prevention systems (IDS/IPS), antivirus, policies and procedures, employee training, and incident response plans.

In addition to the protective measures, the security governance of the industrial infrastructure as a continuous and unstoppable process should also include the assessment and evaluation of the effect of the security measures. Additionally, where it is necessary to eliminate and or minimize vulnerabilities a change should be made. Here, we can also have audit processes that help the security governance process through methods and evaluations of controls that prevent unauthorized access and the review of policies.

We can see and give some concrete examples of security governance measures that can be placed in industrial infrastructure:

- Access controls are presented as a measure that protects us from unauthorized access to information and systems, which have a basis of rights that verify compliance at the time of access. As advanced techniques for the protection of unauthorized access within it, we have, as we know, passwords, two-factor authentication, and biometric authentication.
- Firewalls are appropriate solutions for the security of systems and infrastructures to prevent unauthorized access to networks. They focus on the traffic that enters and leaves the network and the devices that they filter based on some rules defined by the administrator.
- Intrusion detection and prevention systems, known as IDS/IPS systems, enable the detection and prevention of cyber-attacks in the infrastructure. They have almost the same working model as a firewall, except that their greatest power lies in the analysis they do for network traffic for any pattern that could be an attack or breach of security.
- Antivirus is software solutions ready to be installed and operated with antivirus software. They mostly prevent viruses and harmful codes by scanning the system.
- Policies and procedures contain the rules and instructions that must be followed by organizations and staff to maintain the

level of security to protect information and systems. Some standards and frameworks, e.g., ISO 27001, Control Objectives for Information and Related Technologies (COBIT), and NIST propose the use of different policies and procedures that help in the aspect of security governance in organizations.

Some of the examples mentioned above of security governance measures that can be used in industrial infrastructure as a solution to increase the level of security are just a starting point. Industries and organizations must consider security as a component of business operations. Handling or security governance should be done in both technical and non-technical aspects. A comprehensive security program or strategy with an action plan would be appropriate for the industry and its security as well as a continuous review and evaluation of controls and measures.

Although it comes to IT security, exceptional importance is given to ISO 27001, which is the international standard of best practices in the implementation of the information security management system (ISMS). The standard takes care of and helps organizations to protect information and systems from possible threats and maintain confidentiality, integrity, and availability. A point of reference for why we say that ISO 27001 and security governance in industrial infrastructure are related is that it offers an opportunity to have a framework that implements controls and security measures. By following the guidelines and recommendations of ISO 27001, the industry creates a sustainable environment and infrastructure from the aspect of security governance and protects the entire industrial infrastructure from threats (Figure 3.1).

The Control Objectives for Information and Related Technology (COBIT) framework manages and controls organizations' information and technological assets based on a set of best practices that are in line with the organization's goals and objectives. Unlike ISO standards, one way COBIT relates to security governance in industrial infrastructure is that it provides guidance on how to establish controls to protect information and systems. The COBIT framework addresses the aspect of security governance at a broader level and from the higher levels of the organization to the technical ones. Furthermore, in addition to technical control guidelines, COBIT's comprehensive

ISO 27001	ISO 27002
Introduction Scope Normative references Terms and definitions Context of the Organization Leadership Planning Support Operation Performance evaluation Improvement	Introduction Scope Normative references Terms and definitions Organization of information security Human resources security Asset management Access control Cryptography Physical and environmental security Operation security Communications security System acquisition, development and maintenance Supplier relationships Information security incident management Information security aspects of BCM Compliance

Figure 3.1   Content of ISO 27001 and ISO 27002.

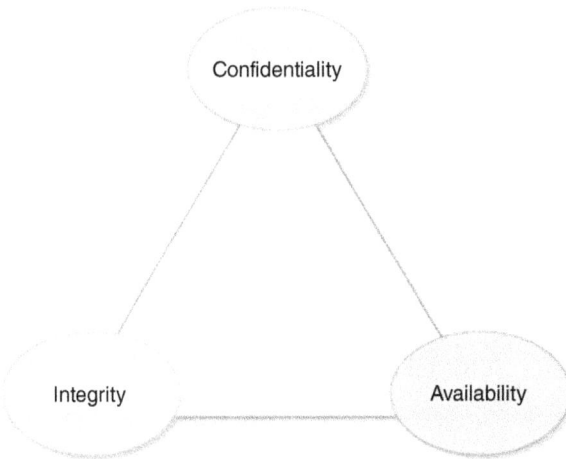

Figure 3.2   COBIT triangle presents a balance between confidentiality, integrity, and availability.

approach also includes non-technical security governance guidelines, including policy development, risk management, and security assessments (Figure 3.2).

From the above discussions and conclusion, we understand that the organization and the industry must advance in the security governance approach within their infrastructures. Including the development along with the technological trends through the utilization of methods, and techniques to achieve the right needed protection of information and systems. The principles, and goals mentioned, and what we know about IT security are now a concern of the entire industry. Since we constantly have threats and advanced attacks in methods and techniques, preventing them has become a problem. To avoid such cases, organizations of all sizes must implement a series of technical and non-technical measures within business operations to protect information and systems from threats such as cyber-attacks, data breaches, and unauthorized access or tampering.

Some crucial measures that can be put in place to provide IT security include technical ones such as access controls, firewalls, intrusion detection and prevention systems (IDS/IPS), and antivirus software. Non-technical includes security awareness programs, policies, and incident management plans. Altogether, in general, it is more important for organizations to protect the confidentiality, integrity, and availability of information and systems from threats and attacks.

Suggested Websites

- International Organization for Standardization (ISO) – ISO/IEC 27001 https://www.iso.org/isoiec-27001-information-security.html
- National Institute of Standards and Technology (NIST) Cybersecurity Framework https://www.nist.gov/cyberframe work
- ISACA – COBIT Framework https://www.isaca.org/resources/cobit

4

THREATS AND ATTACKS TO INDUSTRIAL INTERNET OF THINGS (IIoT)

4.1 Introduction

The Industrial Internet of Things (IIoT) are devices connected to the Internet in industrial systems and processes. They collect data, process it, transfer it to systems, and exchange data with other devices or sensors to perform different tasks. IIoT includes a wide range of devices such as sensors, actuators, and control systems that are used to control and monitor industrial processes. This technology has shown over the years, especially the last ones, that it can make a big difference in terms of efficiency and productivity in the industrial sectors, e.g., production, energy, and oil and refinery.

Nevertheless, technological development and industry innovations are bringing with them new risks, threats, and advanced attacks in methods and techniques that target IIoT. Threats and attacks on IIoT systems are becoming more advanced, and, in most cases, they succeed in targeting and gaining unauthorized access, including cyber-attacks, physical attacks, and insider threats. These unethical interventions disrupt the normality of operations of organizations that provide services, theft of data, and not only the risk of physical damage.

In the continuation of this chapter, we will consider some of the experiences that the industry has gone through and they have left marks on what impact these threats and attacks can have. Well, the security measures can make a difference in the trend of threats and attacks that target IIoT systems. These include some of the most important strong passwords, regularly updating software and firmware, and implementing secure communication protocols. First, a

DOI: 10.1201/9781003383253-4

security awareness program in place for all employees is necessary in these times of numerous technological developments.

4.2 Cyber-Attack Vectors in IoT

Cyber-attacks that target the Internet of Things (IoT) can be carried out from different sides; the most popular attack vectors include:

- Network-based attacks – these types of attacks target the communications in the infrastructure of IoT systems, finding gaps and weaknesses in the network protocols that other devices use to communicate with the world.
- Device-based attacks – these attacks focus on certain devices within the IoT system, e.g., such an attack includes malware infection, firmware modification, and physical tampering.
- Application-based attacks – these attacks focus on the applications used by IoT devices, their weaknesses or gaps within device code or in the interface are the cause of the security breach. Usually, these attacks are carried out through injection attacks, privilege escalation attacks, and unauthorized access.
- Human-based attacks – these attacks are very frequent and have a wider range of action because we cannot understand when they are happening. Social engineering communication techniques are used to deceive people and to compromise the security of IoT systems.

A cyber-attack vector represents a direction which a cyber attacker uses to gain access to computer systems or networks. Subsequently, these cyber-attack vectors take shape, including vulnerabilities, whether in software or hardware, etc. According to the book *Cybersecurity and Cyberwar: What Everyone Needs to Know* by Paul Rosenzweig (Oxford University Press, 2014), *"A vector is simply a way to get from point A to point B. In the case of a cyber-attack, point A is the initial entry of the attacker into the target network, and point B is the place where the attacker wants to be."*. World organizations such as The National Institute of Standards and Technology (NIST) in the framework of cybersecurity present attack vectors that can be included in networks, devices, and applications.

There are various ways that cyber-attackers can attack IoT devices and systems; these include:

- Exploiting vulnerabilities in the device's firmware – this happens in cases where the attacker finds security vulnerabilities and uses them to take control of the firmware of the IoT device. There are many cases when vulnerabilities in the firmware of IoT devices lead to the distribution of well-known distributed denial of service (DDoS) attacks.
- Gaining access through weak or default passwords – password management is a challenge that many organizations neglect. Some IoT devices come from the manufacturer with a default password, and with targeting techniques, they often fall prey to access attacks. The Mirai botnet attacks were the ones that distributed DDoS attacks in 2016 which hacked some IoT devices with default passwords.
- Interception of communication – in these methods of attack, the risk is high and difficult to identify. If the attacker gets access to the communication, there is a possibility of deviation and adaptation and theft of the information that interests him. This can happen between two IoT devices that communicate and transfer data; the attacker can manipulate the data and or gain access to it.
- Physical tampering – these types of attacks are also related to the physical aspect of the IoT device, where physical intervention is attempted for the attacker's purposes of gaining access to that device and the data being transferred.

The organization and the industry in general, should be careful of the manufacturers that sell IoT devices and consider the attack vectors that a specific device may have. Password management, updating software and firmware, and encryption of communications are some protective measures against threats and attacks that can happen very easily. To be protected from these cyber-attacks, organizations must keep all their equipment updated and use strong passwords combined with multi-factor authentication.

4.3 Main Types of Ransomwares

Ransomware is a type of malware that encrypts the victim's data, and will not give access back without fulfilling the demands of the attacker, in this case ransom, and usually, payments are made to restore access to the data. These cyber-attacks are very dangerous if they target entire infrastructures and server systems because they manage to encrypt data and become unusable for the organization. Knowing and recognizing the big attacks of the WannaCry attack in 2017 and the Ryuk attack in 2019, the whole industry is affected. Organizations and world agencies that deal with cybersecurity and critical infrastructures of the industry make continuous efforts to identify and prevent these ransomware attacks.

There are several families, groups, or types of ransomwares, including:

1. Crypto viral extortion – This type of ransomware is typical and common and is used today by attackers. They encrypt the data and demand payment to regain access to it. Examples are Crypto Locker and WannaCry.
2. Locker ransomware – This ransomware blocks access at the level of the operating system, where it does not allow the victim to use his device at all. Ransomware also works here, requesting payment from the victim to unlock access to their device.
3. Scareware – This type of ransomware presents itself as a legitimate software application or even as an antivirus, and the moment it is identified, it demands payment to be removed from the device.
4. Ransomware-as-a-service (RaaS) – This ransomware has found considerable use today because anyone can buy it and use it as a service, even without having the technical knowledge to develop it because there is no need (Table 4.1).

Ransomware attacks in general for the industry pose a great risk and have an impact on their operation and on IIoT systems. Ransomware attacks in IIoT systems have an impact that leaves consequences for the industry that supports day-to-day business operations in these systems. To manage the risks of ransomware attacks on IIoT systems, we must have cybersecurity measures in place that restore the access

Table 4.1 Some Examples of Different Types of Ransomware

TYPE OF RANSOMWARE	DESCRIPTION	EXAMPLE
File-encrypting ransomware	This type of ransomware encrypts files on the victim's device and demands payment in exchange for the decryption key.	WannaCry, Petya, Locky
Master boot record (MBR) ransomware	MBR ransomware infects the master boot record of a victim's computer, preventing the computer from starting up. It then displays a ransom message demanding payment in exchange for restoring the MBR.	Satana, Petya
Mobile ransomware	This type of ransomware targets mobile devices, such as smartphones and tablets, and can encrypt files or lock the device until a ransom is paid.	Android/Filecoder.C, Fusob
Screen-locking ransomware	This type of ransomware locks the victim's device and displays a full-screen message demanding payment to unlock it.	FBI Ransomware, Police Trojan
Ransomware-as-a-service (RaaS)	RaaS is a type of ransomware that allows anyone to become a cybercriminal by renting ransomware from a third-party provider. The provider typically takes a percentage of the ransom payments as their cut.	Satan RaaS, Philadelphia RaaS

back by applying a series of technical and non-technical IT security measures.

There are several examples that we can mention that are related to cyber-attacks in IIoT; in 2017, the WannaCry ransomware attack disrupted the work of several hospitals by encrypting critical data. Another case is that in the 2015 Ukrainian power grid, hackers gained control access to the energy system and caused several losses and power outages in the country.

4.4 Attacks by APTs

APTs (advanced persistent threats) are individuals with many groups or organizations that carry out well-targeted cyber-attacks, continuously and secretly. The focus and targets of APTs remain states, industry, critical infrastructures, and everything else that has a great impact on the nation and society. Through their cyber-attacks, they aim to disrupt social order and cause chaos, leading to life disruptions. APTs are criminal, terrorist organizations, even paid by the state itself, or the army with specific objectives to theft data or disrupt critical infrastructures.

Some of the typical APT attacks include:

- Stuxnet – It was a virus that infected the Iranian nuclear program.
- Operation Aurora – Cyber-attack that targeted technology companies in 2010.
- Equation Group – A group of attackers who were responsible for several APT attacks including the *"EternalBlue"* exploit which was used by the WannaCry ransomware attack.

There have been several cases where APTs have targeted and attacked industrial control systems (ICS) and the IIoT. One of the cases is the example when the group called Dragonfly, which targeted companies in the energy sector in the US and Europe, gained access to the networks of these companies and collected data for their industrial control systems (Figure 4.1).

We will describe the process in several stages of how an APT attack occurs and what impact it has on the industry.

1. Initial compromise – First, the attacker gets access to the target, usually through phishing attacks, and the victims are finding a vulnerability in the system.
2. C2 – Command and control – After gaining access, the attacker aims to control the target, and this phase aims to achieve this from a distance in the compromised system.
3. Lateral movement – The important thing for the attacker is not to lose the initial approach and control he takes on the target. He does not stop and continues in search of further compromise and expansion of the perimeter of the attack.
4. Data exfiltration – This phase has to do with obtaining the data of any information found in the target or victim system.
5. Clearing tracks – The aim of every attacker is to leave no traces at all on the target or victims of the compromised system; this phase has to do with erasing or erasing the traces through the log or audit file.

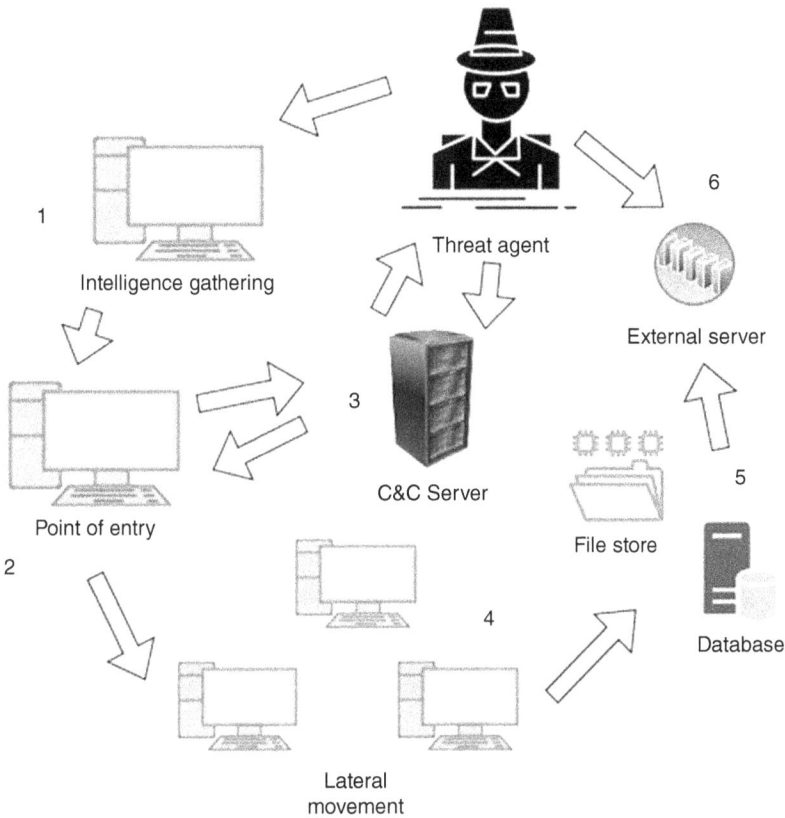

Figure 4.1 Typical stages of an APT attack.

APT attacks are always specified for a specific target and are built for that purpose. They also choose the right time and place for the attack to succeed and for the compromise to be as effective as possible. The mode of action of APT groups makes it difficult to track and predict their further actions and to protect against the attacks they make.

4.5 Statistics

Threats and cyber-attacks in the industry, specifically in the IIoT, have become more frequent and present a serious risk to the industry. The IIoT can be industrial devices that can connect to the Internet. This connection enables remote control of these IIoT devices, which is more productive. But they also present a security risk and vulnerability that attackers can target and exploit against IIoT systems.

Among several threats that IIoT generally faces is the attempt to gain unauthorized access to these devices or the network. Unauthorized access greatly endangers the security of information and the system itself that operates the device with it or through it.

Denial of service (DoS) attacks that target IIoT devices cover the entire device and the network with traffic, making communication with the system difficult. In this way, the user's work is also difficult due to the access they have to the system for monitoring and controlling IIoT devices and the system. In the industry, this presents a problem because it can put equipment and business processes out of order or cause errors.

Malware and physical attacks in IIoT and others are serious threats to IIoT devices and systems, each of which focuses on data and paralyzes the normal operation of industry operations. As we have mentioned in the previous chapters, there are several measures that organizations can take to protect against such threats and attacks. This includes technical and non-technical measures.

Suggested Websites

- Cybersecurity & Infrastructure Security Agency (CISA) https://www.cisa.gov/
- MITRE ATT&CK Framework https://attack.mitre.org/

5

INDUSTRIAL INTERNET OF THINGS (IIoT) VULNERABILITIES

5.1 Introduction to Industrial IoT Vulnerabilities

The Internet of Things (IoT) has seen an immense increase lately, with many interconnected devices being integrated into various industries and critical infrastructures. Industrial sectors, such as manufacturing, oil and refineries, and energy, are seeing the benefits of IoT technologies, such as increased efficiency and improved decision-making processes. Nevertheless, the integration of IoT technologies also brings new risks and threats, potentially security vulnerabilities, which can compromise the triangle of confidentiality, integrity, and availability of industrial systems.

While the number of connected IoT devices and the complexity are increasing, the possibilities for risks, threats, and security breaches from malicious attacks are constantly increasing. These attacks have profound impacts and consequences, including loss of sensitive data, financial loss, and disruption of industry operations. We have experienced several events of these attacks on a global scale, where critical infrastructures have been targeted by cyber-attacks which have caused damage affecting society as well, e.g., the case of Ukrainian power grid blackouts, etc. Such cases of attacks on critical infrastructures with an impact of some size clearly show the importance of the security of IoT systems at industry levels to prevent such attacks by actors who attempt to exploit weaknesses and carry out the attack.

In this chapter, we provide information on the risks, security threats, and vulnerabilities related to Industrial Internet of Things (IIoT) and various approaches and methods of assessment organizations to identify vulnerabilities. We give importance to identifying

DOI: 10.1201/9781003383253-5

those vulnerabilities and securing IIoT systems and the need for an effective countermeasure to protect against cyber-attacks. This chapter provides a base for the discussion of IIoT cybersecurity that will be carried out in the following chapters of this book.

The biggest challenge in securing and protecting IIoT systems and devices is the various range of devices and technologies within this sector involved. These devices include from the very basic sensors to the complex ones, which are used in control systems in oil and refineries or energy plants. This variety of devices and technologies makes it tough to implement in place one cybersecurity solution which is effective for all industrial infrastructure or all systems. Also, the IIoT devices are designed with limited resources in terms of power and memory, which makes it challenging to have robust cybersecurity countermeasures. This situation makes these devices vulnerable to threats, cyber-attacks, and exploitation by hackers or advanced persistent threats APTs.

One more challenge in securing and protecting IIoT systems and devices is the chances for insider threat attacks, which are hard to identify and evidence. Insider threat attacks are related to malicious individuals who have access to the resources and systems, such as employees or third-party contractors. Those malicious individuals, by having this kind of level of access, could bypass security measures and cause serious harm and loss to the systems. Moreover, those insider threats have an understanding and know architecture from the inside of the systems, making it challenging to detect and prevent such threats and attacks.

To mitigate these challenges, organizations must implement a multi-layered approach to cybersecurity measures, including a mix of technical, operational, and organizational security measures. So, the combination of technical and non-technical security measures or controls will give organizations advanced situations in terms of the detection and prevention of such threats and attacks. Furthermore, this approach must employ activities such as a regular assessment of risks that identify vulnerabilities, security audits, checks on secure coding guidelines, and the preparation of response places in place to address any possible security incidents. Also, cybersecurity in the industrial environment of IIoT systems and devices should be designed initially, instead of being added along the way or after.

The integration and utilization of IoT technologies within industrial infrastructures and premises introduce significant security risks and threats that need to focus on addressing the detection and prevention of malicious actors from exploiting vulnerabilities and disrupting critical industrial operations.

5.2 Vulnerability Research for Industrial IoT

As the IoT continues to grow and expand, it is increasingly being integrated into various industrial sectors, including manufacturing, transportation, and energy. In the evolving landscape of the IIoT, the integration of Internet-connected devices into industrial sectors has revolutionized how businesses operate, promising unprecedented levels of efficiency, automation, and data analysis. However, this digital transformation also introduces significant security challenges, with vulnerability research emerging as a critical discipline in safeguarding industrial ecosystems against cyber threats.

5.2.1 Understanding the IIoT Ecosystem

The IIoT ecosystem comprises a complex network of sensors, machines, devices, and software platforms: all interconnected over the Internet. These components work harmoniously to monitor, collect, exchange, and analyze data, facilitating real-time decision-making and operational control. The unique characteristics of the IIoT—such as its scale, heterogeneity, and the critical nature of industrial processes—demand specialized approaches to vulnerability research.

5.2.2 Importance of Vulnerability Research

Vulnerability research in the IIoT context involves systematically identifying, analyzing, and mitigating security flaws within this ecosystem. Given the potential consequences of cyber-attacks on industrial systems—ranging from operational disruption to physical damage and safety hazards—the stakes are exceptionally high. Effective vulnerability research not only protects against external threats but also strengthens the resilience and reliability of industrial operations.

5.2.3 Key Challenges in IIoT Vulnerability Research

1. Complexity and Heterogeneity: The IIoT consists of diverse devices and protocols, many of which have unique vulnerabilities. Researchers must navigate this complexity to understand how different components interact and where security gaps may exist.
2. Real-Time Operational Constraints: Industrial environments often require continuous operation. Conducting vulnerability research without disrupting these operations presents a significant challenge.
3. Legacy Systems Integration: Many industrial setups integrate newer IIoT technologies with legacy systems, which may not have been designed with modern cybersecurity measures in mind. Addressing vulnerabilities in such environments requires innovative approaches that consider both old and new technologies.
4. Supply Chain Security: The IIoT's dependency on a vast network of suppliers and service providers introduces additional vulnerabilities through third-party components. Ensuring the security of these elements is a critical aspect of comprehensive vulnerability research.

5.2.4 Approaches to IIoT Vulnerability Research

1. Automated Vulnerability Scanning: Leveraging automated tools to scan IIoT networks and devices for known vulnerabilities. This approach provides a baseline assessment of the system's security posture.
2. Penetration Testing: Conducting authorized simulated cyber-attacks to identify and exploit weaknesses in the IIoT infrastructure. This hands-on approach helps in understanding the practical implications of potential vulnerabilities.
3. Threat Modeling: Developing models to predict and analyze how attackers might target the IIoT system. Threat modeling assists in identifying security requirements and designing robust defenses.
4. Secure by Design: Incorporating security considerations into the design and development phase of IIoT solutions. This preemptive approach aims to eliminate vulnerabilities at the source.

5. Collaborative Research and Information Sharing: Engaging with the broader cybersecurity community to share knowledge and best practices. Collaboration enhances the collective understanding of IIoT vulnerabilities and defense mechanisms.

The dynamic and complex nature of the IIoT presents significant challenges for vulnerability research. However, by employing a strategic and multifaceted approach, researchers can uncover and address vulnerabilities, thereby fortifying industrial ecosystems against cyber threats. As the IIoT continues to evolve, ongoing research and adaptation to emerging security trends will be paramount in maintaining the integrity and resilience of industrial operations.

5.2.5 Types of IIoT Vulnerabilities

The IIoT encompasses a wide array of technologies, from sensors and devices to networks and data storage: all integrated into industrial processes. As these technologies become increasingly interconnected, they introduce a range of vulnerabilities that can be exploited by cyber attackers. Understanding these vulnerabilities is crucial for implementing effective security measures. Here, we categorize the primary types of IIoT vulnerabilities.

5.2.5.1 Device and Hardware Vulnerabilities

Insecure Interfaces and APIs: Devices may have unprotected interfaces or application programming interfaces (APIs) that allow unauthorized access or control.

Physical Security Breaches: Insufficient physical security measures can lead to unauthorized access, tampering, or damage to devices.

Embedded Systems Weaknesses: Firmware and embedded systems may contain exploitable flaws due to outdated software, lack of secure update mechanisms, or hardcoded passwords.

5.2.5.2 Network Vulnerabilities

Unsecured Wireless Communication: Wireless protocols like WiFi, Bluetooth, and Zigbee, if not properly secured, can be intercepted or disrupted.

Lack of Network Segmentation: Insufficient segmentation of the IIoT network from the corporate network or the Internet can allow attackers to move laterally within systems.

Insecure Data Transmission: Data transmitted across networks without encryption or inadequate encryption can be intercepted, leading to information leakage or manipulation.

5.2.5.3 *Software and Application Vulnerabilities*

Outdated Software: Operating systems, applications, and firmware not regularly updated can contain known vulnerabilities that are easy targets for attackers.

Insufficient Data Validation: Applications that fail to properly validate input data can be susceptible to injection attacks, leading to unauthorized access or execution of malicious code.

Insecure Default Settings: Devices and software shipped with insecure default configurations, such as default passwords or open ports, can provide an easy entry point for attackers.

5.2.5.4 *Authentication and Access Control Vulnerabilities*

Weak Authentication Mechanisms: Systems that rely on weak or easily guessable passwords, or lack multi-factor authentication, can be easily compromised.

Insufficient Access Controls: Lack of proper role-based access control or overly permissive access rights can allow unauthorized users to gain access to sensitive systems or data.

5.2.5.5 *Supply Chain Vulnerabilities*

Third-Party Risks: Dependencies on third-party vendors for hardware, software, or services can introduce vulnerabilities if those parties do not adhere to stringent security practices.

Tampered Components: Hardware components can be intercepted and tampered with before installation, introducing backdoors or malware into the IIoT ecosystem.

5.2.5.6 Environmental and Operational Vulnerabilities

Lack of Situational Awareness: Failure to monitor and respond to changes in the operational environment can leave systems exposed to emerging threats.

Inadequate Incident Response: The absence of a robust incident response plan can exacerbate the impact of security breaches, leading to prolonged disruptions.

The diversity of vulnerabilities in the IIoT landscape necessitates a comprehensive and proactive approach to security. This involves not only technological solutions but also organizational practices such as regular security assessments, employee training, and collaboration with industry partners to share threat intelligence. By understanding the types of vulnerabilities present in IIoT systems, organizations can better prepare and protect their industrial operations from potential cyber threats.

5.2.6 Secondary and Other Types of IIoT Vulnerabilities

In addition to the primary vulnerabilities in the IIoT ecosystem—encompassing device and hardware, network, software and application, authentication and access control, supply chain, and environmental and operational vulnerabilities—there are secondary and other nuanced types that deserve attention. These vulnerabilities can arise from the complex interplay of systems and processes in IIoT environments, often extending beyond straightforward technical flaws to include organizational, procedural, and systemic weaknesses.

5.2.6.1 Configuration and Maintenance Vulnerabilities

Improper Configuration: Misconfigured devices and systems can expose unnecessary risks, such as open ports or services that should not be accessible to the network.

Lack of Regular Maintenance: Failure to regularly maintain and update systems can lead to unpatched vulnerabilities, outdated protocols, and inefficient operations.

5.2.6.2 Protocol and Communication Vulnerabilities

Insecure Legacy Protocols: Many industrial systems rely on legacy protocols that were not designed with security in mind, making them susceptible to eavesdropping, spoofing, and replay attacks.

Insufficient Encryption: Weak or absent encryption for data at rest and in transit can allow attackers to easily intercept and decipher sensitive information.

5.2.6.3 Human Factor Vulnerabilities

Social Engineering: Employees can be manipulated through phishing, pretexting, or other social engineering tactics to gain unauthorized access to IIoT systems.

Insufficient Training: Lack of proper training for staff on cybersecurity best practices and awareness of potential threats can lead to inadvertent security breaches.

5.2.6.4 Environmental Monitoring and Physical Infrastructure Vulnerabilities

Sensor Spoofing: Attackers can manipulate physical environment inputs to sensors, which can lead to incorrect data being reported and potentially causing automated systems to act inappropriately.

Environmental Hazards: Natural disasters, such as floods or earthquakes, can expose physical vulnerabilities in IIoT infrastructure that are not adequately protected against such events.

5.2.6.5 Interoperability and Compatibility Vulnerabilities

System Integration Issues: Challenges in integrating diverse systems and technologies can introduce vulnerabilities, particularly when ensuring secure communication and data sharing between components.

Backward Compatibility: Efforts to maintain backward compatibility with older systems and technologies can sometimes compromise security, leaving systems vulnerable to known exploits.

5.2.6.6 Legal and Regulatory Vulnerabilities

Compliance Risks: Failure to comply with industry standards and regulations can not only result in legal penalties but also indicate potential security weaknesses in systems and practices.

Data Privacy Issues: Inadequate protection of sensitive information, whether related to personnel or operational data, can lead to breaches of privacy laws and regulations.

5.2.6.7 Resource Constraints

Limited Security Investments: Resource constraints, including limited budgets and staffing for cybersecurity initiatives, can result in inadequate security measures and increased vulnerability to attacks.

The secondary and other types of vulnerabilities in IIoT systems underscore the need for a holistic security approach encompassing technical, organizational, and human factors. Addressing these vulnerabilities requires continuous vigilance, regular assessments, and the adoption of a security-first culture within organizations. Collaborative efforts between industry stakeholders and adherence to best practices and standards are essential for mitigating the broad spectrum of risks in the IIoT domain.

As we look toward 2030, the integration of emerging technologies such as artificial intelligence (AI), digital twins, and advanced telecommunications networks like 6G and potentially 7G, into the IIoT will bring about transformative changes. However, these advancements will also introduce new and hazardous vulnerabilities, challenging traditional cybersecurity frameworks and necessitating innovative security solutions. Here are some of the most potentially hazardous vulnerabilities associated with these emerging technologies.

5.2.6.7.1 Artificial Intelligence (AI) and Machine Learning Vulnerabilities (MLV)

Adversarial Attacks: AI systems can be fooled by adversarial inputs designed to be perceived differently by humans and machines, leading to incorrect decisions or actions by IIoT systems.

Data Poisoning: The integrity of AI models can be compromised by injecting malicious data into their training datasets, potentially leading to flawed decision-making processes.

Model Theft: As AI models become more valuable, the risk of intellectual property theft increases, with attackers seeking to reverse-engineer models to exploit vulnerabilities or gain competitive advantages.

5.2.6.7.2 Digital Twins Vulnerabilities

Synchronization Flaws: Discrepancies between the physical assets and their digital counterparts can lead to incorrect analyses and decisions, potentially causing physical damage or operational disruptions.

Privacy and Data Integrity: The extensive data collection and analysis required for digital twins raise significant privacy concerns and vulnerabilities, especially when sensitive operational data is involved.

Access Control and Authentication: Ensuring secure access to digital twins and protecting them from unauthorized changes or manipulations remains a significant challenge.

5.2.6.7.3 Telecommunications Vulnerabilities (6G, 7G, and beyond)

Network Security: Advanced telecommunications technologies will enable faster data transmission and greater connectivity, but they will also increase the attack surface for cyber-attacks, including potential new vectors for large-scale distributed denial of service DDoS attacks.

Ultra-Reliable Low-Latency Communications (URLLC): While URLLC is a boon for critical IIoT applications, it also demands stringent security measures to prevent disruptions that could have immediate and severe consequences.

Quantum Computing Threats: The advent of quantum computing could render current encryption methods obsolete, exposing IIoT systems to eavesdropping and data breaches unless quantum-resistant cryptography is adopted.

5.2.6.7.4 Supply Chain and Third-Party Vulnerabilities

Expanded Attack Surface: As IIoT ecosystems increasingly rely on third-party components and services, including cloud and edge computing resources, the supply chain becomes a more attractive target for attackers seeking to exploit vulnerabilities at any point in the chain.

Interdependence: The interconnectivity and interdependence of services and components in an IIoT ecosystem mean that a vulnerability in one area can have cascading effects, leading to widespread system or network failures.

5.2.6.7.5 Environmental and Physical Security Vulnerabilities

Climate Change Impacts: The increasing frequency of extreme weather events poses physical threats to IIoT infrastructure, including data centers and communication networks, potentially leading to data loss or system outages.

Geopolitical Tensions and Cyber Warfare: IIoT systems, especially those in critical infrastructure, could become prime targets in state-sponsored cyber-attacks, espionage, or sabotage efforts, necessitating robust defenses against sophisticated threats.

The integration of AI, digital twins, and next-generation telecommunications technologies into IIoT will undeniably offer immense benefits. However, it will also introduce complex vulnerabilities that could be exploited in hazardous ways. Addressing these challenges requires forward-thinking cybersecurity strategies that are adaptable, resilient, and capable of evolving with the technological landscape. Organizations must prioritize cybersecurity research, invest in developing secure technologies, and foster collaboration across industries and governments to mitigate these emerging threats effectively.

The interconnection of emerging technologies like AI, digital twins, 6G/7G telecommunications, and their integration into the IIoT significantly amplifies the risk of sophisticated cyber-attacks on industrial infrastructure. This interconnectivity not only broadens the attack surface but also introduces complex vulnerabilities that can be exploited in coordinated and high-impact cyber-attacks.

5.2.7 *Amplified Attack Surface*

Integrated Systems: The seamless integration of AI, digital twins, and advanced telecommunications with IIoT devices creates a tightly interconnected ecosystem. While this enhances operational efficiency and data flow, it also means that a single vulnerability can provide a gateway for attackers to compromise multiple parts of the infrastructure.

5.2.8 *Cross-Technology Exploits*

Exploiting AI Weaknesses: Attackers could use AI to analyze data from various IIoT devices and systems, identifying patterns and vulnerabilities to exploit. Conversely, compromising AI systems could lead to incorrect decisions affecting the physical operations of critical infrastructure.

Digital Twins as Targets: Digital Twins, which simulate real-world industrial processes, could be manipulated to disrupt physical operations. By feeding false information into these simulations, attackers could cause industrial systems to operate under dangerous or damaging conditions without immediate detection.

Advanced Telecommunications Exploits: The adoption of 6G and 7G technologies promises to increase the speed and volume of data transfer, enabling more dynamic and responsive IIoT applications. However, these networks also introduce new vulnerabilities, such as the potential for more sophisticated denial-of-service attacks that could cripple critical infrastructure.

5.2.9 *Coordinated Multi-vector Attacks*

Complex Cyber-Physical Attacks: The convergence of cyber and physical systems in the IIoT means that cyber-attacks can have direct physical consequences. For example, attackers could simultaneously target AI-driven decision-making algorithms and the digital twins modeling critical infrastructure processes, causing widespread operational chaos.

Supply Chain Compromise: An attack on one part of the supply chain can have cascading effects throughout the IIoT ecosystem. Attackers could exploit interconnected vulnerabilities, using compromised components or software as a conduit to disrupt critical infrastructure operations.

5.2.10 Data Integrity and Privacy Concerns

Data Manipulation: The integrity of data flowing through IIoT systems is paramount. Attackers could manipulate data at any point in its journey—from collection by sensors and transmission over 6G/7G networks to analysis by AI algorithms—leading to erroneous outcomes with potentially hazardous implications.

Privacy Breaches: The vast amount of data collected and analyzed by IIoT systems, including potentially sensitive operational information, makes privacy breaches a significant risk. Such breaches could expose proprietary information or critical infrastructure vulnerabilities to malicious actors.

5.2.11 Mitigation Strategies

As IIoT ecosystems evolve, understanding and mitigating the interconnected risks will be crucial for protecting critical industrial infrastructure from sophisticated and high-risk cyber-attacks. To counter these risks, it's essential to adopt a multi-layered security approach that includes:

- End-to-end encryption across all data transmission points to protect data integrity and confidentiality.
- Robust authentication and access control mechanisms to ensure that only authorized entities can interact with IIoT systems and data.
- Regular security assessments to identify and mitigate vulnerabilities, including penetration testing and red team exercises.
- Resilience planning to prepare for and respond to cyber-attacks, ensuring that critical operations can continue even when parts of the IIoT ecosystem are compromised.
- Collaboration and information sharing among industry stakeholders, security experts, and government agencies to stay ahead of emerging threats.

5.2.12 Methods of Vulnerability Assessment

Vulnerability assessment is a critical process in managing cybersecurity risks, especially in complex environments like the IIoT. It involves identifying, quantifying, and prioritizing (or ranking) vulnerabilities in a system. Here are several key methods of vulnerability assessment that organizations can employ to safeguard their IIoT infrastructure:

1. Automated Vulnerability Scanning
   Automated vulnerability scanners are tools that scan systems, networks, and applications for known vulnerabilities. These tools use databases of known vulnerabilities, such as the common vulnerabilities and exposures (CVE) list, to identify potential security issues. They can be configured to scan regularly, providing ongoing visibility into the security posture of an organization's IIoT environment.

2. Penetration Testing
   Penetration testing (pen testing) is a proactive and simulated cyber-attack against your system to check for exploitable vulnerabilities. In the context of IIoT, pen testing can help identify weaknesses in devices, applications, and network protocols. Unlike automated scanning, penetration testing is usually performed manually by experts who can think like attackers and potentially uncover complex exploit chains that automated tools might miss.

3. Security Audits and Reviews
   Security audits and reviews involve a comprehensive examination of an organization's information security system. This method includes reviewing policies, access controls, and physical security, as well as examining how security protocols are implemented within IIoT devices and infrastructure. It's a more holistic approach that looks beyond just technical vulnerabilities to understand the organizational and procedural factors that could impact security.

4. Wireless Security Assessment
   Given the extensive use of wireless technologies in the IIoT, conducting specific assessments on wireless networks is crucial. This involves evaluating the security of WiFi, Bluetooth, Zigbee, and other wireless protocols used within the IIoT

ecosystem. Wireless security assessments look for vulnerabilities that could allow unauthorized access, eavesdropping, or data tampering.

5. Configuration and Patch Management Assessment

   This method focuses on evaluating the current state of device and software configurations across the IIoT environment, ensuring they are optimized for security. It also involves assessing how patches and updates are managed to ensure that systems are protected against known vulnerabilities. Effective configuration and patch management are vital for maintaining the security and integrity of IIoT devices and systems.

6. Threat Modeling

   Threat modeling is a process by which potential threats, such as structural vulnerabilities or the absence of appropriate safeguards, can be identified and mitigations can be planned. In the context of IIoT, threat modeling helps to anticipate the ways an attacker might exploit the system and allows for the development of strategies to mitigate these risks before they can be exploited.

7. Supply Chain Risk Assessment

   Assessing the security of the supply chain is critical in IIoT environments, where components and software often come from a variety of vendors. This method involves evaluating the security practices of suppliers, the integrity of components, and the potential for security weaknesses introduced through third-party elements.

8. Physical Security Assessment

   Physical security assessments are crucial for IIoT devices that may be deployed in inaccessible or remote locations. This involves evaluating the physical access controls to devices and systems, assessing the risk of tampering, theft, or physical damage, and ensuring that adequate protective measures are in place.

A comprehensive vulnerability assessment strategy for IIoT should employ a blend of these methods to ensure a thorough evaluation of risks. By regularly conducting vulnerability assessments, organizations can significantly reduce the attack surface of their IIoT environments and strengthen their overall cybersecurity posture.

Below is a matrix that outlines key components of such a strategy, mapping assessment methods to key cybersecurity areas in the IIoT ecosystem. Table 5.1 serves as a guide for organizations to enhance their cybersecurity posture by ensuring a broad and effective approach to identifying and mitigating vulnerabilities.

Implementation notes:

- **Automated Vulnerability Scanning**: Regularly scan all components for known vulnerabilities, especially focusing on device and network security. Automate scans to run at frequent intervals.
- **Penetration Testing**: Conduct annual or bi-annual penetration tests to identify vulnerabilities in devices, networks, software, and wireless protocols that automated tools might miss.
- **Security Audits and Reviews**: Perform comprehensive reviews of security policies, procedures, and the implementation of security controls across the IIoT ecosystem. Include device security, network security, and physical infrastructure in the audit scope.
- **Wireless Security Assessment**: Assess the security of wireless communications used by IIoT devices. Focus on encryption standards, authentication protocols, and the resilience of wireless networks against eavesdropping or data tampering.
- **Configuration and Patch Management Assessment**: Ensure devices, software, and systems are correctly configured and regularly updated. Assess the effectiveness of current patch management practices to mitigate vulnerabilities promptly.
- **Threat Modeling**: Develop threat models for the IIoT ecosystem, identifying potential threats to device security, data security, network security, and wireless protocols. Use these models to guide the development of security measures.
- **Supply Chain Risk Assessment**: Evaluate the security of the supply chain, focusing on the integrity of devices and software. Assess the security practices of suppliers and the risk of introducing vulnerabilities into the IIoT ecosystem.
- **Physical Security Assessment**: Assess the physical security controls in place to protect IIoT devices and infrastructure. This includes evaluating access controls, environmental controls, and the risk of physical tampering.

Table 5.1 Overview of Integration Cybersecurity Area with the Security Measures in the Organization

CYBERSECURITY AREA	AUTOMATED VULNERABILITY SCANNING	PENETRATION TESTING	SECURITY AUDITS AND REVIEWS	WIRELESS SECURITY ASSESSMENT	CONFIGURATION AND PATCH MANAGEMENT ASSESSMENT	THREAT MODELING	SUPPLY CHAIN RISK ASSESSMENT	PHYSICAL SECURITY ASSESSMENT
Device security	✓	✓	✓	✓	✓	✓	✓	✓
Network security	✓	✓	✓	✓	✓	✓		
Software security	✓	✓	✓		✓	✓	✓	
Data security	✓	✓	✓	✓	✓	✓	✓	
Access control	✓	✓	✓		✓	✓		✓
Wireless protocols		✓	✓	✓		✓		
Physical infrastructure								✓

This matrix is a strategic framework designed to ensure a holistic approach to vulnerability assessment in the IIoT. By addressing each area comprehensively, organizations can significantly enhance the cybersecurity of their IIoT sensors and systems, reducing the risk of cyber-attacks and ensuring the resilience of critical industrial processes.

5.2.13 Current State of Vulnerability Research

The current state of vulnerability research reflects a rapidly evolving field driven by the continuous emergence of new technologies, the increasing complexity of cyber-attacks, and the ever-expanding digital landscape. This dynamic environment presents both challenges and opportunities for researchers, security professionals, and organizations across the globe. Here are some key trends and considerations in the field.

5.2.14 Increasing Focus on Automation and AI

The use of automation and artificial intelligence (AI) in vulnerability research has become more prevalent. These technologies can significantly enhance the efficiency and effectiveness of vulnerability discovery, analysis, and mitigation. AI is being used to predict potential vulnerabilities based on patterns and to automate the scanning of code for flaws. However, this also raises concerns about AI-powered attacks and the need for AI-secure coding practices.

5.2.15 Cloud and Supply Chain Vulnerabilities

With the growing reliance on cloud services and third-party suppliers, vulnerabilities in these areas have become a major focus. The complexity of supply chains and the shared responsibility model of cloud services introduce multiple points of potential exposure. Researchers are increasingly focused on identifying and mitigating risks associated with third-party components, APIs, and cloud configurations.

5.2.16 IoT and OT Security

The expansion of the IoT and the integration of operational technology (OT) with IT systems have opened new fronts for vulnerability research. The heterogeneity of devices, protocols, and standards, along with their widespread use in critical infrastructure, necessitates targeted research to secure these environments against sophisticated threats.

5.2.17 Zero-Day Vulnerabilities

The discovery and disclosure of zero-day vulnerabilities—previously unknown vulnerabilities exploited by attackers before developers can issue patches—remain at the forefront of vulnerability research. The race between researchers and attackers to identify these vulnerabilities underscores the need for proactive security measures and rapid response mechanisms.

5.2.18 Quantum Computing and Cryptographic Vulnerabilities

As quantum computing advances, researchers are examining the potential impact on cryptographic standards and security protocols. Quantum-resistant cryptography is becoming a crucial area of study to prepare for a future where traditional encryption methods may no longer be secure.

The current state of vulnerability research is characterized by rapid technological changes, the increasing interconnectedness of systems, and the constant evolution of cyber threats. This environment demands continuous adaptation, innovation, and collaboration among researchers, security professionals, and organizations to protect against vulnerabilities and ensure the resilience of digital systems and infrastructures.

5.3 Cybersecurity and Industrial IoT Trends

The IIoT represents a significant transformation in industrial and manufacturing processes, introducing a new level of efficiency, automation, and data analysis. However, this digital integration also brings about critical cybersecurity challenges. As we navigate through the evolving landscape of IIoT, understanding current trends in cybersecurity is essential for safeguarding these advanced systems. Here are some important trends in cybersecurity and IIoT.

5.3.1 Increasing Integration of AI and Machine Learning

- Automated Threat Detection and Response: The adoption of artificial intelligence (AI) and machine learning (ML) in cybersecurity tools enables more sophisticated detection of anomalies and potential threats in real time, enhancing the protection of IIoT systems.
- Predictive Security Measures: AI and ML are increasingly used to predict vulnerabilities and potential attack vectors, allowing organizations to implement proactive defenses before incidents occur.

5.3.2 Enhanced Focus on Edge Security

- Securing Edge Devices: As IIoT expands, edge computing devices proliferate, processing data closer to its source. Securing these devices becomes paramount, as they often operate outside traditional security perimeters.
- Robust Authentication and Encryption: Ensuring the integrity and confidentiality of data processed at the edge involves implementing stronger authentication mechanisms and end-to-end encryption.

5.3.3 Adoption of Zero-Trust Architectures

- Micro-segmentation and Least Privilege: The zero-trust model, which assumes that threats can originate from anywhere, is becoming more relevant for IIoT environments. This approach emphasizes the micro-segmentation of networks and enforcing least privilege access control to minimize the attack surface.
- Continuous Monitoring and Verification: Continuous verification of all devices and users within the IIoT ecosystem ensures that security is not compromised, aligning with the zero-trust principle of "never trust, always verify."

5.3.4 Supply Chain Security

- Third-Party Risk Management: The complex supply chains involved in IIoT highlight the need for rigorous security evaluations of third-party vendors and components, preventing vulnerabilities from being introduced into the system.

- Secure Software Development Lifecycle (SDLC): Emphasizing security within the SDLC of IIoT devices and software helps mitigate risks from the outset, including those related to the supply chain.

5.3.5 Regulatory Compliance and Standards

- Adhering to Industry Standards: As IIoT grows, compliance with industry standards and regulations (such as National Institute of Standards and Technology NIST frameworks, ISO/IEC 27001, and sector-specific guidelines) becomes crucial for maintaining cybersecurity and ensuring interoperability.
- Global Data Protection Regulations: With the increasing amount of data generated by NIST devices, adherence to data protection laws (e.g., General Data Protection Regulation GDPR, California Consumer Privacy Act CCPA) is essential for maintaining consumer trust and legal compliance.

5.3.6 Emphasis on Resilience and Incident Response

- Building Resilient Systems: Designing IIoT systems with resilience in mind allows them to maintain operations even when under attack, minimizing potential disruptions.
- Enhanced Incident Response Plans: Organizations are developing comprehensive incident response plans tailored to the unique aspects of IIoT environments, ensuring rapid recovery and mitigation of damages in the event of a breach.

5.3.7 Awareness and Training

- Cybersecurity Training: As human error remains a significant vulnerability, there is an increased focus on cybersecurity awareness and training for all personnel involved in the operation and management of IIoT systems.

The trends in cybersecurity and IIoT underscore a shift toward more integrated, proactive, and resilient approaches to security. As technologies evolve and cyber threats become more sophisticated, staying informed about these trends is critical for protecting the industrial ecosystem. By adopting advanced security technologies, embracing comprehensive risk management strategies, and ensuring compliance with regulatory standards, organizations can navigate the complexities of IIoT cybersecurity effectively.

5.4 Penetration Testing Approach and Methods

Penetration testing, often referred to as "pen testing" or ethical hacking, is a critical cybersecurity practice where simulated cyber-attacks are carried out on a computer system, network, or web application to identify vulnerabilities that could be exploited by malicious actors. This process involves a combination of manual and automated techniques to systematically compromise servers, endpoints, web applications, wireless networks, network devices, mobile devices, and other potential points of exposure. Here, we explore the approach and methods involved in penetration testing within the context of enhancing security, particularly in the IIoT environments.

5.4.1 Phases of Penetration Testing

- Planning and Reconnaissance
    - Objective Definition: Clearly define the scope and goals of the penetration test, including the systems to be addressed and the testing methods to be used.
    - Intelligence Gathering: Collect as much information as possible about the target environment to identify potential entry points and vulnerabilities. This includes public domain information, network and system details, and application functionalities.
- Scanning
    - Static Analysis: Inspect code to estimate how it behaves while running. This can help to identify immediate vulnerabilities.
    - Dynamic Analysis: Assess how the application behaves during execution to discover vulnerabilities that appear only during operation.

- Network and Port Scanning: Tools like Nmap are used to scan the system's network to identify open ports and running services.
- Gaining Access
  - Utilize web application attacks such as cross-site scripting, SQL injection, and backdoors to uncover vulnerabilities.
  - Techniques such as brute-forcing, exploit execution, and social engineering attacks are used to establish access.
- Maintaining Access
  - Aim to understand if the vulnerability can be used to achieve a persistent presence within the exploited system, simulating APTs that remain in the system for months to exfiltrate or compromise sensitive data.
- Analysis
  - Analyze the data gathered during the test to identify vulnerabilities, security holes, and risks. This phase results in a report detailing the vulnerabilities, the sensitive data that was accessed, the amount of time the pen tester was able to remain undetected in the system, and recommendations for mitigation.

5.4.2 Penetration Testing Methods

- Black Box Testing
  The tester has no prior knowledge of the network infrastructure. This simulates an attack by an external hacker and helps to understand how an attacker approaches a target.
- White Box Testing
  The tester has full knowledge of the network and system infrastructure. This comprehensive testing is useful for thorough vulnerability assessments.
- Gray Box Testing
  It is a combination of both black and white box testing, where limited knowledge about the system is shared with the tester. It offers a balance between the depth of the test and the time required to perform it.

5.4.3 Special Considerations for IIoT

- Device Diversity: IIoT environments are characterized by a wide range of devices with varying operating systems and protocols, requiring specialized knowledge for effective penetration testing.
- Operational Technology (OT) Integration: Pen tests must consider the integration of IT and OT environments, recognizing the potential for physical damage or safety hazards if operational systems are disrupted.
- Regulatory Compliance: IIoT environments often operate under strict regulatory requirements, making compliance a key consideration during pen testing.
- Real-Time Data: Many IIoT applications involve real-time data processing, necessitating tests that account for the potential impact on system performance and data integrity.

Penetration testing is a vital element in the cybersecurity strategy for IIoT environments, providing critical insights into potential vulnerabilities and the effectiveness of existing security measures. By following a structured approach and applying appropriate testing methods, organizations can significantly enhance their security posture, safeguarding against the evolving landscape of cyber threats.

5.5 OSINT Approach for Industrial IoT Vulnerability Search

Open-source intelligence (OSINT) plays a pivotal role in the cybersecurity domain, especially in identifying vulnerabilities within the IIoT ecosystem. OSINT involves collecting and analyzing information from publicly available sources to support intelligence needs, including the identification of potential vulnerabilities in hardware, software, and network configurations used within industrial environments. This section outlines the OSINT approach specifically tailored for discovering vulnerabilities in IIoT systems.

5.5.1 Steps in OSINT Approach for IIoT Vulnerability Search

- Define Objectives and Scope
  - Identify the goals of the OSINT research, focusing on specific components, devices, or systems within the IIoT

ecosystem. Determine the scope to ensure targeted and efficient information gathering.

- Gathering Information
  - Public Databases and Repositories: Utilize databases such as the National Vulnerability Database (NVD), Common Vulnerabilities and Exposures (CVE) list, and other security advisories to find known vulnerabilities affecting IIoT devices and software.
  - Search Engines and Specialized Tools: Use search engines and tools like Shodan, Censys, or ZoomEye to uncover exposed IIoT devices and systems, analyze their footprint, and identify potentially vulnerable endpoints.
  - Social Media and Forums: Monitor discussions on platforms like Twitter, Reddit, and specialized forums where researchers and practitioners might share insights about vulnerabilities, exploits, and security incidents.
  - Technical Documentation and Datasheets: Review publicly available technical documentation, datasheets, and user manuals for IIoT devices to understand their architecture, default configurations, and known issues.
- Analysis and Correlation
  - Analyze the gathered information to identify patterns, correlations, and actionable intelligence. This includes mapping discovered vulnerabilities to specific devices or systems, assessing their potential impact, and identifying any existing exploits.
  - Employ tools and frameworks for automating the analysis process, enabling the efficient handling of large datasets and the extraction of relevant insights.
- Validation
  - Validate the identified vulnerabilities against actual devices or systems, when possible, to confirm their existence and exploitable nature. This step may involve the use of penetration testing techniques under controlled conditions.
- Reporting and Mitigation
  - Document the findings in a comprehensive report detailing the identified vulnerabilities, their potential impact, and recommended mitigation strategies.

- Share the report with relevant stakeholders, including security teams, device manufacturers, and software developers, to initiate the remediation process.

5.5.2 Best Practices and Considerations

- Ethical and Legal Compliance: Ensure that all OSINT activities comply with relevant laws and ethical guidelines, respecting privacy and avoiding unauthorized access to systems.
- Continuous Monitoring: Adopt a proactive stance by continuously monitoring for new information and threats, as the landscape of vulnerabilities in the IIoT domain is constantly evolving.
- Collaboration and Information Sharing: Engage with the cybersecurity community, industry groups, and governmental agencies to share findings and contribute to collective knowledge bases.

The OSINT approach for IIoT vulnerability search is a critical component of a comprehensive cybersecurity strategy, enabling organizations to proactively identify and address potential threats. By leveraging publicly available information, organizations can enhance their understanding of the threat landscape, prioritize security efforts, and strengthen the resilience of their IIoT ecosystems against cyber threats.

Suggested Websites

- OWASP IoT Project https://owasp.org/www-project-internet-of-things/
- MITRE ATT&CK Framework for ICS https://attack.mitre.org/
- Cybersecurity & Infrastructure Security Agency (CISA)—ICS Security https://www.cisa.gov/ics

6

REFERENCES FRAMEWORKS, STANDARDS, AND OTHER REGULATIONS

In the domain of cybersecurity, particularly within the context of the Industrial Internet of Things (IIoT), the outlook is continually changed by diverse frameworks, standards, and regulations. These guidelines not only aim to establish best practices for securing systems and data but also serve to standardize ways over industries, ensuring interoperability, reliability, and trust. This chapter provides an overview of the essential references, frameworks, standards, and regulations that play pivotal roles in governing the cybersecurity posture of IIoT systems globally.

Cybersecurity frameworks (CSFs) offer structured methodologies for managing and mitigating cyber risks. They are instrumental in guiding organizations through the complexities of establishing robust cybersecurity practices, especially in environments as nuanced and diverse as IIoT.

- NIST CSF: Developed by the National Institute of Standards and Technology, the NIST CSF provides a policy framework of computer security guidance for how private sector organizations in the United States can assess and improve their ability to prevent, detect, and respond to cyber-attacks.
- ISA/IEC 62443: This series of standards, developed by the International Society of Automation (ISA) and adopted by the International Electrotechnical Commission (IEC), focuses on the secure development, maintenance, and operation of Industrial Automation and Control Systems (IACS).

- ISO/IEC 27001: Part of the ISO/IEC 27000 family of standards, ISO/IEC 27001 specifies the requirements for establishing, implementing, maintaining, and continually improving an information security management system (ISMS).

Standards are important for ensuring that devices, systems, and processes adhere to recognized security benchmarks, facilitating compatibility and interoperability across the IIoT ecosystem.

- IEEE Standards: The Institute of Electrical and Electronics Engineers publishes various standards that impact IIoT, including those focused on network communication, data integrity, and privacy.
- IEC Standards for Industrial Automation: Beyond ISA/IEC 62443, the IEC publishes numerous standards that cover different aspects of industrial automation, including safety, communications, and system integration.

Regulations are legally binding rules that organizations must follow, often varying significantly by geography and industry sector. Compliance with these regulations is critical to avoid legal penalties and protect the organization's reputation.

- General Data Protection Regulation (GDPR): A regulation in EU law on data protection and privacy in the European Union and the European Economic Area (EEA), GDPR also addresses the transfer of personal data outside the EU and EEA areas.
- California Consumer Privacy Act (CCPA): This act enhances privacy rights and consumer protection for residents of California, United States.
- Network and Information Systems (NIS) Directive: The EU's directive on the security of network and information systems aims to raise levels of security and resilience of network and information systems across the EU.

As IIoT continues to emerge, so too will the frameworks, standards, and regulations that govern it. Emerging trends such as the integration of blockchain for enhanced security, the use of artificial intelligence for predictive threat analysis, and the development of quantum-resistant cryptographic standards will likely impact future updates and the creation of new guidelines. The integration of frameworks, standards, and regulations into the fabric of IIoT cybersecurity strategies is indispensable. These guidelines not only provide a roadmap for securing intricate systems but also promote a culture of continuous advancement and adaptation in the face of evolving cyber threats. As the IIoT landscape continues to expand, staying informed and compliant with these guidelines will be paramount for organizations striving to protect their assets and maintain trust with stakeholders and customers alike.

6.1 Introduction to Regulations and Compliance in IIoT

In the increasingly evolving world of the IIoT, the convergence of operational technology (OT) with information technology (IT) brings forth unprecedented levels of efficiency, productivity, and innovation. However, this integration also introduces complex cybersecurity challenges, as the interconnected nature of IIoT systems exposes critical infrastructure to cyber threats. To mitigate these risks, a robust framework of regulations and compliance standards is essential. The IIoT ecosystem comprises a wide range of devices, from sensors and actuators to complex control systems, all of which collect and exchange vast amounts of data. This data is not only valuable for operational efficiency but also critical for safety and privacy. As such, ensuring the security and integrity of these systems is paramount. Regulations and compliance standards are developed to provide a structured approach to cybersecurity, offering guidelines that help protect against data breaches, unauthorized access, and other cyber threats.

Compliance plays a pivotal role in maintaining trust in IIoT systems. By adhering to established regulations and standards, organizations can demonstrate their commitment to cybersecurity, data protection, and privacy. Compliance helps in:

- Establishing security best practices where regulations often encapsulate the collective wisdom of cybersecurity experts and offer organizations a roadmap to securing their systems.
- Enhancing customer trust compliance assures customers and partners that the organization takes security seriously, which is crucial for building and maintaining trust.

- Mitigating legal and financial risks, non-compliance can result in hefty fines and legal penalties, not to mention the reputational damage associated with data breaches.

Several regulations have a straight impact on the security and operation of IIoT systems, as follows:

- General Data Protection Regulation (GDPR): It affects how personal data collected by IIoT devices is handled, emphasizing the need for consent and the right to privacy.
- California Consumer Privacy Act (CCPA): It is like GDPR but focused on the rights of California residents, imposing requirements on businesses collecting personal data.
- Network and Information Systems (NIS) Directive targets the resilience of network and information systems, which is crucial for the reliable operation of IIoT in critical infrastructure.
- Sector-Specific Regulations: Various industries, such as healthcare, energy, and manufacturing, may also be subject to additional regulations that address specific risks and requirements.

Achieving compliance in the IIoT domain is fraught with challenges, including:

- Complexity and Diversity: The wide variety of devices and systems, each with its vulnerabilities and requirements, complicates the compliance process.
- Evolving Regulations: The regulatory landscape is continually changing, requiring organizations to stay informed and adapt their compliance strategies accordingly.
- Global Operations: Multinational organizations must navigate a maze of regulations that vary by country and region.

As we deep dive further into the specifics of various regulations, standards, and compliance strategies in the following sections, it becomes clear that regulation and compliance form the backbone of a secure and trustworthy IIoT ecosystem.

6.2 NIST Cybersecurity Framework in the Context of IIoT

The National Institute of Standards and Technology (NIST) CSF is a framework primarily considered for critical infrastructure organizations to manage and mitigate cybersecurity risk. However, its principles and practices apply to industries, including the IIoT. This framework provides a policy structure that organizations can adapt to protect their networks and information from cyber threats. Within the IIoT context, the NIST CSF offers a thorough approach to enhancing security across interconnected systems and devices.

6.2.1 Overview of the NIST CSF

The NIST CSF is structured around five core functions, which are further broken down into categories and subcategories detailing specific outcomes and security controls. These five core functions are identify, protect, detect, respond, and recover. Each function plays a crucial role in creating a holistic cybersecurity strategy, particularly relevant in the complex ecosystems of IIoT.

- Identify: This function involves developing an organizational understanding of managing cybersecurity risk to systems, assets, data, and capabilities. For IIoT, this means mapping out the network of devices and understanding their role and potential vulnerabilities.
- Protect: This function outlines safeguards to ensure the delivery of critical services. In IIoT, this could involve securing data in transit and at rest, managing access to IIoT devices, and ensuring that devices and software are regularly updated.
- Detect: This involves the development and implementation of appropriate activities to identify the occurrence of a cybersecurity event. IIoT systems must be monitored in real-time to detect anomalies or unauthorized access attempts.
- Respond: Once a cybersecurity event is detected, this function focuses on acting regarding a detected cybersecurity incident. This could involve isolating affected devices, conducting forensic analysis, and implementing procedures to maintain operations.
- Recover: This function involves plans for resilience and recovery from a cybersecurity incident. For IIoT, recovery plans must consider not just data recovery but also the restoration of operational capabilities, which may be critical to safety and production processes.

6.2.2 Application of the NIST Framework in IIoT

Implementing the NIST CSF in IIoT environments requires consideration of the unique characteristics and challenges associated with these systems, including their heterogeneity, real-time processing needs, and potential impact on physical processes. Key considerations include:

- Comprehensive Asset Management: Given the diverse range of devices in IIoT, identifying and managing assets is critical. This includes understanding the lifecycle of each device, from deployment to decommissioning.
- Tailored Protection Measures: Protection strategies must account for the operational context of IIoT devices, balancing security with the need for uninterrupted service. For example, patching devices in an industrial control system requires careful planning to avoid disrupting production.
- Enhanced Detection Capabilities: IIoT systems often generate vast amounts of data, necessitating advanced analytical tools to detect threats effectively. Machine learning and anomaly detection techniques can provide early warning of potential issues.
- Incident Response for Operational Continuity: Responding to incidents in IIoT environments must prioritize operational continuity, ensuring that essential functions can be maintained even when parts of the network are compromised.
- Recovery and Resilience Planning: Recovery strategies should include not only data and system restoration but also plans for alternative operational modes that can be employed in the event of a cyberattack.

The NIST CSF offers a valuable structure for managing cybersecurity risks in IIoT environments. By adapting and implementing its core functions, organizations can enhance their security posture while maintaining the operational integrity of their IIoT systems. As the IIoT landscape continues to evolve, leveraging frameworks such as NIST ensures a robust and flexible approach to cybersecurity, essential for the protection of critical infrastructure and industrial processes.

6.3 CISA Regulations in the Context of Industrial IoT

The Cybersecurity and Infrastructure Security Agency (CISA) plays a crucial role in advancing the security of the nation's critical infrastructure, with a focus that extends into the realm of the IIoT. As part of the United States Department of Homeland Security, CISA provides guidance, tools, and resources to support the implementation of robust cybersecurity measures across various sectors. Overview of CISA's Role and Regulations

CISA was established to strengthen the cybersecurity posture and resilience of national critical infrastructure. It achieves this through collaboration with partners in both the public and private sectors, providing expertise, insights, and resources to counteract cyber threats. For IIoT, CISA's regulations and guidance play a significant role in shaping security practices, particularly in sectors such as energy, manufacturing, and critical manufacturing.

6.3.1 Key CISA Initiatives Impacting IIoT

- National Cybersecurity Protection System (NCPS): Also known as EINSTEIN, this integrated system provides capabilities to detect, prevent, and respond to cyber threats. IIoT environments can benefit from the insights and alerts generated by NCPS, enabling timely responses to potential security incidents.
- Industrial Control Systems (ICS) Cybersecurity: CISA provides a wealth of resources and tools specifically designed for securing ICS, which are integral to IIoT. This includes the ICS Cybersecurity Advisory Program, which offers vulnerability assessments, incident response services, and training to enhance the security of industrial control systems.
- The National Risk Management Center (NRMC): Focused on identifying and addressing significant risks to critical infrastructure, the NRMC collaborates with industry partners to develop strategies and initiatives that bolster the resilience of critical systems, including those in the IIoT domain.
- Cybersecurity Assessments: CISA offers cybersecurity assessment tools and services, such as the Cyber Resilience Review

(CRR) and the Vulnerability Scanning service, which organizations can utilize to evaluate their cybersecurity practices and identify areas for improvement.

- Information Sharing and Collaboration: Through the Automated Indicator Sharing (AIS) initiative and the Critical Infrastructure Information Sharing and Collaboration Program (CIISCP), CISA facilitates the exchange of threat intelligence and best practices among stakeholders, enhancing the collective security posture of IIoT ecosystems.

While CISA's directives and guidelines are not always mandatory, adhering to them is considered best practice for organizations operating within or supporting critical infrastructure sectors. Compliance involves:

- Conducting Regular Assessments: Utilizing CISA's tools and services to evaluate and enhance cybersecurity measures.
- Implementing Recommended Controls: Applying security controls and practices recommended by CISA to protect IIoT environments.
- Engaging in Information Sharing: Participating in information-sharing initiatives to stay informed about emerging threats and mitigation strategies.
- Collaborating with CISA: Engaging with CISA during incidents or when seeking guidance on securing IIoT systems and networks.

Implementing CISA's guidance within IIoT environments may present challenges, including the need to balance operational efficiency with security and the complexity of securing a diverse and distributed network of devices. Organizations must also navigate the evolving regulatory landscape, ensuring compliance with all applicable standards and guidelines. CISA's regulations and initiatives provide a critical foundation for securing the IIoT, offering resources, expertise, and a framework for enhancing cybersecurity measures. By aligning with CISA's guidance, organizations can better protect their critical infrastructure from cyber threats, ensuring the resilience and reliability of their IIoT systems. As the cybersecurity landscape continues to evolve, the role of CISA in supporting and securing IIoT environments will remain indispensable, highlighting the importance of ongoing collaboration and compliance.

6.4 ISO/IEC 27000 Series: Securing the Industrial Internet of Things (IIoT)

The ISO/IEC 27000 series introduces a suite of information security standards published jointly by the International Organization for Standardization (ISO) and the International Electrotechnical Commission (IEC). These standards provide a framework for managing and securing information assets, offering a comprehensive approach to information security that is applicable across various sectors, including the rapidly evolving domain of the IIoT. This section explores the relevance and application of the ISO/IEC 27000 series within the context of IIoT, highlighting key standards and their role in enhancing cybersecurity measures for interconnected industrial systems.

The ISO/IEC 27000 series, also known as the "ISMS Family of Standards," encompasses guidelines and best practices for implementing, maintaining, and improving an ISMS. The series addresses various aspects of information security management, including risk management, security controls, and compliance. Among these standards, ISO/IEC 27001 is the most widely recognized, providing requirements for establishing, implementing, operating, monitoring, reviewing, maintaining, and continually improving an ISMS.

6.4.1 Key Standards Relevant to IIoT

- ISO/IEC 27001—Information Security Management: It establishes requirements for creating and maintaining an ISMS, crucial for protecting data integrity, confidentiality, and availability in IIoT environments.
- ISO/IEC 27002—Code of Practice for Information Security Controls: It offers best practice recommendations on information security controls under the guidance of the principles laid out in ISO/IEC 27001, applicable to securing IIoT devices and data.

- ISO/IEC 27005—Information Security Risk Management: It provides guidelines for managing information security risks in a manner consistent with ISO/IEC 27001, essential for assessing and mitigating risks in IIoT systems.
- ISO/IEC 27017—Code of Practice for Information Security Controls Based on ISO/IEC 27002 for Cloud Services: Addresses cloud security aspects relevant to IIoT platforms and services hosted in cloud environments.

The adoption and application of the ISO/IEC 27000 series within IIoT contexts involve several key considerations:

- Risk Assessment and Management: Conduct comprehensive risk assessments for IIoT systems, identifying potential threats to information security and implementing appropriate controls based on ISO/IEC 27005 guidance.
- Data Protection and Privacy: Ensure the confidentiality, integrity, and availability of data collected, processed, and stored by IIoT devices, in line with ISO/IEC 27001 requirements.
- Device and Network Security: Apply ISO/IEC 27002 controls to secure IIoT devices and their communications, protecting against unauthorized access, data breaches, and other cyber threats.
- Cloud Security: For IIoT solutions leveraging cloud services, follow ISO/IEC 27017 guidelines to address specific security challenges associated with cloud computing.

6.4.2 Benefits of ISO/IEC 27000 Compliance in IIoT

- Enhanced Security Posture: Implementing the ISO/IEC 27000 series standards helps organizations strengthen their security measures, reducing the risk of cyber incidents.
- Improved Regulatory Compliance: Compliance with these standards demonstrates adherence to recognized information security practices, facilitating regulatory compliance and customer trust.
- Systematic Risk Management: The structured approach to risk management advocated by the ISO/IEC 27000 series enables organizations to identify, assess, and mitigate risks effectively.

- Operational Resilience: Establishing robust information security management practices contributes to the resilience of IIoT systems, ensuring operational continuity even in the face of security challenges.

The ISO/IEC 27000 series provides a valuable framework for managing information security in the context of IIoT. By adopting and adapting these standards, organizations can enhance the security and resilience of their IIoT ecosystems, protecting critical industrial processes and sensitive data against cyber threats. As IIoT continues to advance, aligning with the principles and practices outlined in the ISO/IEC 27000 series will remain essential for safeguarding interconnected industrial environments.

6.5 OWASP Open Community Methodology: Enhancing IIoT Security

The Open Web Application Security Project (OWASP) is renowned for its contributions to improving software security across various domains, including the IIoT. OWASP's open community methodology, which leverages the collective knowledge and expertise of cybersecurity professionals worldwide, has proven invaluable in identifying, understanding, and mitigating security risks in software development and deployment.

OWASP is an international nonprofit organization focused on improving the security of software through open-source projects, research, standards, and educational resources. Its open community approach encourages participation and collaboration among security experts, developers, and other stakeholders to share knowledge and develop free, publicly available security tools and resources.

While OWASP's primary focus has been on web application security, its principles, methodologies, and tools are increasingly relevant to the IIoT domain, where software plays a critical role in device operation, data management, and system integration. Key OWASP projects and resources applicable to IIoT security include:

- OWASP Internet of Things Project: This initiative provides a comprehensive framework for understanding and addressing security issues specific to the Internet of Things (IoT) and, by extension, IIoT. It offers practical guidance, including the

IoT Top Ten, a list of the top ten security concerns for IoT developers and manufacturers.

- OWASP Top Ten: Although primarily targeting web applications, the OWASP Top Ten list of critical security risks serves as a valuable reference for IIoT software security, highlighting areas such as injection flaws, authentication issues, and data exposure that are equally pertinent to IIoT environments.
- OWASP Software Assurance Maturity Model (SAMM): SAMM offers an effective framework for assessing and improving software security practices within an organization. IIoT stakeholders can utilize SAMM to integrate security considerations throughout the software development lifecycle, from design and implementation to deployment and maintenance.
- OWASP Security Knowledge Framework (SKF): SKF is an open-source web application that provides developers with knowledge and tools to build secure applications, including code examples and documentation on secure coding practices. Applying SKF's guidelines can help mitigate common security vulnerabilities in IIoT software components.

Implementing OWASP's open community methodology within IIoT involves several strategic approaches:

- Community Engagement and Collaboration: Encourage participation in OWASP projects and events to stay abreast of the latest security research, tools, and best practices that can be applied to IIoT.
- Adopting OWASP Guidelines: Integrate OWASP's security guidelines, such as those from the IoT Project and the Top Ten, into IIoT development and deployment processes to address prevalent security risks.
- Leveraging OWASP Tools: Utilize OWASP's tools and resources for vulnerability assessment, code review, and security testing in IIoT systems to identify and remediate potential security flaws.
- Contributing to OWASP Projects: Organizations and individuals involved in IIoT can contribute their expertise and experiences back to the OWASP community, enriching the collective knowledge base and fostering the development of new security solutions.

The OWASP open community methodology offers a dynamic and collaborative approach to tackling the complex security challenges faced by the IIoT ecosystem. By embracing OWASP's principles, leveraging its resources, and engaging with its community, IIoT stakeholders can significantly enhance the security and resilience of their systems. As the IIoT landscape continues to evolve, the open exchange of knowledge and the development of community-driven security solutions will be key to safeguarding interconnected industrial environments against emerging cyber threats.

6.6 OSSTMM Methodology: A Comprehensive Approach to IIoT Security

The Open-Source Security Testing Methodology Manual (OSSTMM) is a rigorous framework developed by the Institute for Security and Open Methodologies (ISECOM) for conducting in-depth security tests and assessments. While originally focused on information systems and networks, the principles and practices outlined in the OSSTMM can be effectively applied to the IIoT to ensure a thorough evaluation of security risks and vulnerabilities. The OSSTMM provides a comprehensive guide for performing security tests across five key areas: information and data controls, personnel security, physical security, wireless communications, and network security. By focusing on these areas, the OSSTMM aims to provide a complete picture of an organization's security posture, identifying potential vulnerabilities and offering actionable insights for improvement. The methodology emphasizes a scientific approach to security testing, relying on verifiable facts to assess and quantify security risks.

Applying the OSSTMM methodology to IIoT involves a holistic examination of the various components that constitute IIoT systems, including devices, networks, software applications, and human factors. The following outlines how the OSSTMM can be adapted to address the unique security challenges of IIoT:

- Information and Data Controls: Assess the methods used to protect data collected, processed, and stored by IIoT devices, focusing on encryption, access controls, and data integrity mechanisms.
- Personnel Security: Evaluate the security awareness and practices of individuals who interact with IIoT systems, including

employees, contractors, and third-party service providers. This involves examining training programs, access privileges, and adherence to security policies.

- Physical Security: Analyze the physical safeguards in place to protect IIoT devices and infrastructure from unauthorized access, tampering, or theft. This includes assessing the security of manufacturing facilities, data centers, and other locations where IIoT components are deployed.
- Wireless Communications: Investigate the security of wireless protocols and networks used for communication between IIoT devices. This involves testing for vulnerabilities in WiFi, Bluetooth, Zigbee, and other wireless technologies.
- Network Security: Conduct a thorough examination of the network infrastructure supporting IIoT operations, including firewalls, routers, and other network devices. This includes assessing network segmentation, intrusion detection systems, and other mechanisms used to protect the network perimeter and internal communications.

6.6.1 Key Benefits of Applying OSSTMM to IIoT

- Comprehensive Risk Assessment: The OSSTMM provides a framework for conducting a thorough evaluation of security risks across multiple dimensions, offering a holistic view of an organization's security posture.
- Actionable Insights: By quantifying security risks and identifying specific vulnerabilities, the OSSTMM methodology delivers actionable insights that organizations can use to strengthen their IIoT security measures.
- Improved Security Planning: The detailed approach outlined in the OSSTMM supports more effective security planning and resource allocation, ensuring that efforts are focused on addressing the most critical vulnerabilities.
- Enhanced Compliance: Applying the OSSTMM methodology can help organizations meet regulatory requirements and industry standards related to cybersecurity, enhancing overall compliance posture.

The OSSTMM methodology offers a structured and comprehensive approach to evaluating and improving the security of IIoT systems. By adapting and applying its principles to the unique context of IIoT, organizations can achieve a deeper understanding of their security risks and implement more effective measures to protect their critical infrastructure. As the IIoT landscape continues to grow and evolve, leveraging established methodologies like the OSSTMM will be essential for maintaining robust security practices and safeguarding against emerging threats.

6.7 PTES Methodology: Enhancing Security in Industrial IoT

The Penetration Testing Execution Standard (PTES) is a comprehensive framework that outlines the phases and methodologies involved in conducting a penetration test. Developed to standardize the process of penetration testing and ensure a thorough examination of a system's security, PTES provides a detailed approach that can be particularly beneficial in the context of the IIoT.

PTES is divided into seven main sections, each designed to guide security professionals through the critical stages of a penetration test, from initial engagement to final reporting. The standard covers:

- Pre-engagement Interactions: Establishing the scope, objectives, and legal considerations of the penetration test.
- Intelligence Gathering: Collecting as much information as possible about the target system, including publicly available data and more targeted reconnaissance.
- Threat Modeling: Identifying potential threats and vulnerabilities within the system to focus the penetration test on areas of greatest risk.
- Vulnerability Analysis: Systematically identifying exploitable vulnerabilities in the system using a combination of automated tools and manual techniques.
- Exploitation: Attempting to exploit identified vulnerabilities to determine their impact and the level of access that can be achieved.
- Post Exploitation: Determining the value of the machine compromised and the potential further access it can provide within the target environment.

• Reporting: Documenting the findings, methodologies used, and evidence gathered during the test, along with recommendations for remediation and improvement.

Applying the PTES methodology to IIoT systems involves a tailored approach that considers the unique aspects of industrial environments, such as operational continuity, safety requirements, and the heterogeneity of devices and protocols.

• Pre-engagement Interactions: Given the critical nature of IIoT systems, defining clear boundaries and objectives while ensuring minimal disruption to operational processes is essential.
• Intelligence Gathering: This phase can include mapping the IIoT network, identifying connected devices, and understanding communication flows and data processing activities.
• Threat Modeling: Special attention should be given to threats that could impact the physical operations of the IIoT, such as sabotage, espionage, and safety breaches.
• Vulnerability Analysis: This involves not only scanning for known vulnerabilities in software and firmware but also assessing proprietary protocols and custom-built IIoT solutions for potential weaknesses.
• Exploitation: Exploitation attempts in IIoT environments must be conducted with caution, considering the potential for causing physical damage or disrupting critical operations.
• Post Exploitation: Understanding the impact of a successful exploit includes evaluating the potential for lateral movement within the IIoT environment and access to sensitive operational data.
• Reporting: Reports should provide actionable recommendations that balance security improvements with operational requirements, ensuring that remediation efforts do not adversely affect IIoT functionality.

6.7.1 Key Benefits of PTES in IIoT

• Comprehensive Security Assessment: The structured approach of PTES ensures a thorough assessment of IIoT systems, identifying vulnerabilities across both IT and operational technology (OT) components.

- Focused Remediation Efforts: By prioritizing vulnerabilities based on their potential impact, PTES helps organizations allocate resources effectively to address the most critical security issues.
- Enhanced Understanding of Attack Surfaces: PTES aids in developing a detailed understanding of the attack surfaces within IIoT environments, facilitating better protection strategies against targeted attacks.
- Improved Compliance and Risk Management: The thorough documentation and reporting process outlined in PTES supports compliance with regulatory requirements and enhances overall risk management strategies.

The PTES provides a robust framework for conducting penetration tests that can be effectively applied to the security challenges of IIoT systems. By following the PTES methodology, organizations can gain deeper insights into their IIoT security posture, identify and mitigate vulnerabilities, and enhance the resilience of their industrial operations against cyber threats. As the complexity and connectivity of IIoT environments continue to increase, adopting standardized methodologies like PTES will be crucial for maintaining a strong cybersecurity defense.

6.8 Open IIoT: Embracing Openness in Industrial IoT Development and Security

The concept of "Open IIoT" refers to the application of open-source principles, standards, and technologies in the development, deployment, and security of IIoT systems. This approach leverages the collaborative, transparent, and community-driven nature of open-source projects to enhance innovation, interoperability, and security within the IIoT ecosystem.

Open IIoT integrates open-source software, hardware, and standards into the fabric of industrial automation and connectivity. It promotes the use of openly available resources and community-supported projects to build and secure IIoT solutions, contrasting with proprietary systems that may rely on closed-source technologies. Key components of Open IIoT include:

- Open-Source Software and Platforms: Utilizing open-source operating systems, middleware, and applications for IIoT devices and systems.
- Open Hardware: Adopting open specifications for physical components and devices, enabling easier customization and integration.
- Open Standards: Following open communication protocols and data formats to ensure interoperability and data exchange among diverse IIoT devices and systems.

6.8.1 Advantages of Open IIoT

- Enhanced Security: Open IIoT benefits from the open-source model's inherent transparency, allowing for broader scrutiny of code and designs by the global community, leading to the identification and remediation of vulnerabilities more efficiently.
- Increased Interoperability: By adhering to open standards, Open IIoT facilitates seamless integration between different devices, systems, and platforms, reducing vendor lock-in and enabling more flexible and scalable solutions.
- Cost-Effectiveness: Open-source solutions can reduce the costs associated with licensing fees and proprietary technologies, making IIoT deployments more accessible to a wider range of organizations.
- Rapid Innovation: The collaborative nature of open-source projects accelerates innovation, as developers and engineers worldwide contribute improvements, features, and fixes, driving the rapid evolution of IIoT technologies.

6.8.2 Challenges and Considerations

While Open IIoT presents numerous benefits, several challenges must be addressed to realize its full potential:

- Quality Assurance and Support: Ensuring the reliability and stability of open-source components in critical industrial applications requires robust testing and support frameworks, which may not always be readily available.
- Security Management: The open nature of the software and hardware components demands diligent security practices, including regular patching and updates, to protect against vulnerabilities.
- Integration Complexity: Achieving seamless integration between open-source and existing proprietary systems may require significant customization and technical expertise.
- Compliance and Intellectual Property: Navigating the legal implications of open-source licenses and ensuring compliance with industry regulations and standards is crucial for organizations adopting Open IIoT.

Fostering the growth of Open IIoT involves several key strategies:

- Community Engagement: Actively participating in open-source communities and projects related to IIoT, contributing to development, testing, and documentation efforts.
- Adoption of Open Standards: Advocating for and implementing open standards in IIoT projects to promote interoperability and vendor neutrality.
- Education and Awareness: Raising awareness about the benefits and challenges of Open IIoT among stakeholders, including industry leaders, policymakers, and educators, to encourage broader adoption.
- Collaborative Security Initiatives: Engaging in collaborative security initiatives, such as shared vulnerability databases and joint response teams, to enhance the collective security posture of the Open IIoT ecosystem.

Open IIoT represents a paradigm shift toward a more collaborative, transparent, and innovative approach to industrial connectivity and automation. By leveraging the strengths of open-source software, open hardware, and open standards, organizations can achieve more secure, interoperable, and cost-effective IIoT solutions. Despite the challenges, the ongoing commitment of the global community to address these issues ensures that Open IIoT remains a viable and promising path for the future of industrial operations.

6.9 FIRST Guides: Strengthening Incident Response in Industrial IoT

The Forum of Incident Response and Security Teams (FIRST) is an international confederation of trusted computer incident response teams that cooperatively handle computer security incidents and promote incident prevention programs. FIRST publishes a range of guides, standards, and best practices designed to facilitate the effective management and resolution of security incidents. In the context of the IIoT, the FIRST guides provide invaluable resources for developing robust incident response capabilities tailored to the unique challenges of interconnected industrial environments.

FIRST guides cover various aspects of incident response, from initial setup and preparation to managing the technical and organizational response to an incident. These guides are developed by experienced security professionals and are intended to be applicable across different sectors and technologies, including the emerging domain of IIoT. Some key guides and standards provided by FIRST include:

- Computer Security Incident Response Team (CSIRT) Services Framework: A framework that outlines the range of services a CSIRT might offer, providing a foundational structure for incident response teams.
- Traffic Light Protocol (TLP): A set of guidelines designed to facilitate greater sharing of sensitive information, simply and standardized across different contexts and communities.
- Simulation Exercise Guidelines: Recommendations for planning, conducting, and evaluating simulation exercises to test and improve incident response capabilities.

The application of FIRST guides within IIoT environments involves adapting and implementing their principles to address the specific operational, safety, and reliability requirements of industrial systems:

- Building Specialized IIoT CSIRTs: Establish dedicated incident response teams for IIoT environments, utilizing the CSIRT Services Framework to define their roles, responsibilities, and services. These teams should possess expertise in both IT security and operational technology (OT) to effectively address incidents in IIoT contexts.

- Enhanced Information Sharing: Apply the traffic light protocol to facilitate secure and efficient sharing of threat intelligence and incident information among IIoT stakeholders, including manufacturers, operators, and sector-specific ISACs (information sharing and analysis centers).

- Conducting IIoT-focused Simulation Exercises: Use FIRST's Simulation Exercise Guidelines to design and execute drills that simulate realistic IIoT security incidents, helping to identify potential weaknesses in incident response plans and improve team readiness.

- Developing Incident Response Plans: Leverage FIRST's best practices to develop comprehensive incident response plans that address the full lifecycle of an incident, from detection and analysis to containment, eradication, recovery, and post-incident activities.

6.9.1 Benefits of Implementing FIRST Guides in IIoT

- Improved Preparedness: FIRST guides help organizations develop a proactive stance toward incident response, enhancing their ability to detect and respond to incidents quickly and effectively.

- Strengthened Resilience: By establishing clear procedures and facilitating better communication and collaboration, FIRST guides contribute to the overall resilience of IIoT systems against cyber threats.

- Knowledge Sharing and Collaboration: Adopting FIRST's standardized practices for information sharing can improve collaboration within and across industries, leading to a stronger collective defense against emerging security challenges.

- Enhanced Recovery Capabilities: Effective incident response planning and simulation exercises ensure that IIoT environments can recover more rapidly from security incidents, minimizing operational disruptions and potential safety risks.

FIRST guides offer a comprehensive set of resources for enhancing incident response capabilities within the IIoT sector. By adopting and adapting these guides to the specific needs of IIoT environments, organizations can strengthen their ability to manage and respond to security incidents, thereby safeguarding critical industrial processes and infrastructures against cyber threats. As the IIoT landscape continues to evolve, the principles and practices outlined in FIRST guides will remain integral to developing resilient and responsive security strategies.

6.10 Best Practices to Secure Industrial IoT Devices

The proliferation of IIoT devices has brought about significant advancements in operational efficiency, productivity, and innovation within various industrial sectors. However, the integration of these interconnected devices into critical infrastructure also presents substantial cybersecurity risks. Ensuring the security of IIoT devices is paramount to protect against unauthorized access, data breaches, and potential disruptions to critical operations.

6.10.1 Secure Device Configuration

- Change Default Credentials: Always change default usernames and passwords to strong, unique credentials to prevent unauthorized access.
- Disable Unnecessary Services: Turn off any services or features not required for the device's operation to minimize potential attack vectors.
- Regular Updates and Patch Management: Ensure that firmware and software are regularly updated to address known vulnerabilities. Implement a systematic patch management process to apply updates promptly.

6.10.2 Network Segmentation and Access Control

- Implement Network Segmentation: Separate IIoT devices from the rest of the network using firewalls or virtual LANs (VLANs) to limit the spread of potential attacks.
- Use Access Control Lists (ACLs): Define and enforce policies that restrict which devices and users can communicate with IIoT devices, ensuring only authorized access.

6.10.3 Data Protection and Privacy

- Encrypt Data: Use strong encryption for data at rest and in transit to protect sensitive information from interception or tampering.
- Secure Data Storage: Ensure that data stored on IIoT devices or associated storage media is protected with encryption and access controls.

6.10.4 Secure Communication Protocols

- Use Secure Protocols: Opt for secure communication protocols, such as TLS/SSL, for encrypting data in transit. Avoid protocols that do not support encryption or are known to be vulnerable.
- Implement Mutual Authentication: Utilize mutual authentication mechanisms to ensure that both the IIoT device and the communicating entity are authenticated.

6.10.5 Continuous Monitoring and Anomaly Detection

- Deploy Intrusion Detection Systems (IDS): Use IDS tailored for industrial networks to monitor for suspicious activities or anomalies indicative of a cyberattack.
- Regular Security Audits: Conduct periodic security audits of IIoT devices and systems to identify potential vulnerabilities or misconfigurations.

6.10.6 Vulnerability Management

- Conduct Vulnerability Assessments: Regularly assess IIoT devices for vulnerabilities using automated tools and manual testing techniques.
- Stay Informed on Threat Intelligence: Subscribe to threat intelligence feeds and security advisories relevant to IIoT devices to stay aware of emerging threats and vulnerabilities.

6.10.7 Incident Response Planning

- Develop an Incident Response Plan: Prepare a detailed incident response plan tailored to the IIoT environment, outlining procedures for detecting, containing, eradicating, and recovering from security incidents.
- Train Staff: Ensure that all relevant personnel are trained on the incident response plan and their specific roles during a security incident.

6.10.8 Secure Development Practices

- Adopt a Secure Development Lifecycle (SDLC): For organizations developing IIoT solutions, integrating security into every stage of the development process is crucial to identifying and mitigating risks early.
- Security by Design: Design IIoT devices with security as a fundamental consideration, incorporating features like secure boot, hardware-based trust anchors, and the ability to securely update firmware.

Securing IIoT devices requires a comprehensive and multi-layered approach that addresses both the technological and organizational aspects of cybersecurity. By implementing these best practices, organizations can significantly enhance the security posture of their IIoT environments, protecting critical industrial systems and data against cyber threats. As the IIoT landscape continues to evolve, maintaining vigilance and adapting to new security challenges will be essential for safeguarding the future of industrial operations.

Suggested Websites

- **ISO/IEC 27001—Information Security Management** https://www.iso.org/isoiec-27001-information-security.html
- National Institute of Standards and Technology (NIST) https://www.nist.gov/
- European Union Agency for Cybersecurity (ENISA) https://www.enisa.europa.eu/

7

THE APPROACH PROPOSED
BY REFERENCES

7.1 Introduction

The increase of the Industrial Internet of Things (IIoT) is a transformative time in the industrial sector, indicating unseen levels of connectivity, efficiency, and automation. Yet, this digital transformation is not without its challenges, particularly in the realm of cybersecurity. The integration of IIoT devices into critical industrial processes makes these systems vulnerable to cyber threats, from data breaches to operational disruptions. In response, a lot of cybersecurity references, including frameworks, standards, and guidelines, have come out, providing structured approaches to safeguard IIoT ecosystems.

The core of this chapter is to anatomize the approaches proposed by these cybersecurity references, making their guidance into actionable strategies for securing IIoT systems. By deepening into the basics of security principles, understanding the structured frameworks, acknowledging the crucial cybersecurity requirements, and implementing robust countermeasures, organizations can steer the intricacies of IIoT security with assurance.

Adopting cybersecurity references serves multiple purposes. Primarily, it offers organizations an assessment blueprint for constructing and advancing their cybersecurity posture. These references refine the collective knowledge and expertise of cybersecurity professionals worldwide, providing best practices that have been refined through real-world application. Furthermore, obedience to recognized standards and frameworks can ease regulatory compliance, advance stakeholder trust, and set up a common language for discussing and addressing cybersecurity issues within the IIoT domain.

DOI: 10.1201/9781003383253-7

This chapter centers on offering a detailed investigation of the security fundamentals key to protecting IIoT systems. It starts by setting up a thorough understanding of the security challenges these systems face and builds a foundation for the application of relevant cybersecurity frameworks (CSFs) and standards. The discussion then deeps into the structure of essential references, investigating the architecture of leading CSFs and standards and illuminating their components as well as their applicability to IIoT settings. Furthermore, this chapter emphasizes important cybersecurity requirements, identifying the core points that must be addressed to secure IIoT systems effectively, with insights from reliable references. Lastly, it explores practical cybersecurity countermeasures, introducing actionable strategies specially made to mitigate the risks related to IIoT, ensuring a solid and resilient security posture for those systems.

Through the investigation of these objectives, this chapter tries to equip organizations with the knowledge and tools needed to defend their IIoT systems against cyber threats and make certain the resilience and integrity of their industrial operations in the digital age.

7.2 Security Fundamentals for Industrial IoT

The surfacing of the IIoT introduces a significant change in how industrial operations are conducted, monitored, and optimized. This integration of digital technologies with physical industrial processes, while providing a multitude of benefits in terms of efficiency and innovation, also presents an intricate array of cybersecurity challenges. Understanding the security fundamentals related to IIoT is important for organizations to protect their operations against cyber threats effectively.

IIoT systems are distinguished by their integration of various devices and technologies, including sensors, actuators, and software, into industrial processes. Apart from traditional IT settings, IIoT ecosystems not only manage data but also have direct control over most physical operations. This dual nature notably elevates the stakes for cybersecurity, as breaches can result in not just data loss or theft but also physical damage and safety risks. The unique features of IIoT security introduce clear challenges that must be addressed to ensure solid protection. Operational endurance is a critical concern, as many industrial processes are important and can't afford disruptions, making them durable against cyber threats. Additionally, IIoT devices often relate directly to the physical world, controlling machinery and processes, which presents remarkable safety and physical security risks if these systems are compromised or theft. The variousness of IIoT settings further intricate security, as they typically are mixed with modern and legacy systems, each introducing unique vulnerabilities and challenges. Moreover, the wide attack surface created by the vast number of devices and the intricacy of IIoT networks provides many entry points for attackers, underlining the need for thorough security countermeasures.

To address the challenges of securing IIoT systems, key principles underpin effective cybersecurity strategies. Defense in depth commands the application of multiple security layers throughout the IIoT ecosystem, making sure that the compromise of one layer does not endanger the complete system. The concept of least privilege limits access to IIoT systems and data, allowing only the vital access needed for individuals and systems to operate, thus diminishing the potential impact of breaches. Security by design highlights the importance of incorporating security considerations from the very beginning of the design and development phases for IIoT devices and systems rather than considering security later. Moreover, regular monitoring and assessment are crucial for detecting threats and vulnerabilities proactively through ongoing system surveillance and routine security assessments. Lastly, a solid incident response and recovery plan is key for rapidly containing, eliminating, and recovering from security incidents, thereby minimizing their effects on operations and safety. Together, these principles draw up a comprehensive framework for protecting IIoT settings.

Securing IIoT systems needs addressing a number of built-in challenges that come from their intricacy and dynamism. Issues of scalability and management arise due to the airy number of devices, making it hard to deploy and maintain effective security countermeasures over the entire network. Furthermore, interoperability adds another layer of difficulty, as the need for smooth communication over diverse devices and systems often creates vulnerabilities, particularly when integrating legacy systems with modern technologies. Protecting data privacy and integrity is also important, as

IIoT devices frequently collect and transmit sensitive data that needs protecting, particularly in industries with demanding regulatory standards. In addition, the dimension of emerging threats asks for ongoing attention and adaptability as attackers continuously advance their tactics, techniques, and procedures to exploit weaknesses in IIoT settings. To effectively resist these challenges, there is a need for innovative and robust security strategies customized to the unique features of IIoT systems.

Implementing effective IIoT security countermeasures needs a thorough strategy that addresses the challenges and complies with basic security principles. Network segmentation is a vital measure of security, involving the separation of networks into smaller, manageable segments to limit the spread of attacks' lateral movement and minimize the exposure of attack surface. Strong authentication and encryption are important for securing communications and protecting sensitive data, making sure that only authorized entities have access. Ongoing vulnerability assessments play a vital role in maintaining a strong cybersecurity posture by identifying and addressing weaknesses in the system. Additionally, promoting a culture of security awareness is key, as educating staff and stakeholders about cybersecurity best practices and the related risks associated with IIoT settings helps build a proactive and informed security mindset. These strategies work together to offer a solid framework for protecting IIoT environments against emerging threats.

Securing IIoT systems needs a detailed understanding of the distinct security challenges and principles that support effective cybersecurity strategies. By embracing a thorough approach that includes defense in depth, regular monitoring, and a dedication to security by design, organizations can steer the intricacy of IIoT security and protect their critical industrial processes against cyber threats. As IIoT continues to emerge, staying informed and adapting to new security developments will be vital to maintaining robust defenses in this dynamic outlook.

7.3 Structure of the References

In the outlook of IIoT cybersecurity, various crucial references serve as the backbone for developing, implementing, and managing effective security countermeasures. These references arch over frameworks, standards, and guidelines providing orderly approaches to protect IIoT systems against emerging cyber threats.

7.3.1 NIST Cybersecurity Framework (CSF)

The National Institute of Standards and Technology (NIST) CSF is a vital reference designed to guide organizations in managing and minimizing cybersecurity risks. The framework is arranged around five core functions that constitute the lifecycle of an effective cybersecurity program, as follows:

- Identify: Establish a foundational understanding of managing cybersecurity risk to systems, assets, data, and capabilities.
- Protect: Implement appropriate safeguards to ensure the delivery of critical services.
- Detect: Develop and employ the necessary activities to identify the occurrence of a cybersecurity event.
- Respond: Act regarding a detected cybersecurity incident.
- Recover: Maintain plans for resilience and restore any capabilities or services impaired due to a cybersecurity incident.

Each basic function is divided into categories and subcategories, defining specific outcomes and referencing informative resources, making the CSF adaptable to organizations of all sizes and sectors, including IIoT.

7.3.2 ISO/IEC 27000 Series

The ISO/IEC 27000 series consists of information security standards that offer a model for building, implementing, operating, monitoring, reviewing, maintaining, and improving an Information Security Management System (ISMS). The structure of the series includes as follows:

- ISO/IEC 27001: Specifies the requirements for an ISMS, focusing on a risk management process that is integral for protecting IIoT devices and data.
- ISO/IEC 27002: Provides a code of practice for information security controls, detailing specific measures that can be applied to secure IIoT environments.

- ISO/IEC 27005: Focuses on information security risk management, a critical aspect of securing IIoT systems by identifying, assessing, and treating risks.

The series promotes a structured and ongoing way of managing information security risks that are especially suited to the dynamic nature of IIoT ecosystems.

7.3.3 OWASP Guidelines

The Open Web Application Security Project (OWASP) is renowned for its contributions to improving software security. Although initially focused on web applications, OWASP's guidelines are increasingly relevant to IIoT. Key resources include:

- OWASP Top Ten: Lists the ten most critical web application security risks, which can be applied to IIoT software components to identify and mitigate vulnerabilities.
- OWASP IoT Project: Offers comprehensive guidance specifically tailored to Internet of Things (IoT) devices, including those used in IIoT settings, focusing on secure development, deployment, and maintenance.

OWASP guidelines are structured to provide actionable recommendations that developers and security professionals can implement to enhance the security of IIoT applications and devices.

7.3.4 IEC 62443 Series

The International Electrotechnical Commission (IEC) 62443 series of standards is designed to secure Industrial Automation and Control Systems (IACS). It provides a flexible framework to address and mitigate current and future security vulnerabilities in IACS. The series is structured into several sections that cover general policies and procedures, system, component, and product development lifecycle requirements, offering a comprehensive approach to securing all aspects of IACS, which form the backbone of IIoT.

7.3.5 PTES Methodology

The Penetration Testing Execution Standard (PTES) offers a detailed methodology for conducting penetration tests, structured into seven main sections:

- Pre-engagement interactions define the scope and objectives of the penetration test.
- Intelligence gathering involves collecting information about the target system or network.
- Threat modeling identifies potential threats and vulnerabilities.
- Vulnerability analysis systematically identifies exploitable vulnerabilities.
- Exploitation attempts to exploit identified vulnerabilities.
- Post exploitation assesses the value of compromised systems.
- Reporting documents findings and recommendations.

PTES provides a structured approach that can be particularly beneficial for identifying and mitigating vulnerabilities within IIoT systems.

The structure of these references incorporates a thorough and multi-faceted approach to cybersecurity, fitting out to the specific needs and challenges of IIoT settings. By understanding and benefiting from the components and methodologies sketched out in these references, organizations can develop robust security strategies that protect against cyber threats while enabling the continued innovation and operational efficiency that IIoT promises. As the IIoT landscape evolves, so too will these references, adapting to new challenges and incorporating emerging best practices to guide organizations in securing their critical industrial systems.

7.4 Essential Requirements for IIoT Cybersecurity

The IIoT surrounds many interconnected devices and systems that are integral to the operations of various industry sectors, including manufacturing, energy, and transportation. This alike, while enabling unseen levels of efficiency and automation, also presents notable cybersecurity challenges. To protect against these challenges, it is vital to adhere to a set of essential cybersecurity requirements tailored to the needs of IIoT systems.

7.4.1 Robust access control and authentication mechanisms

Given the likely critical nature of IIoT devices and the data they handle, implementing strong access control and authentication mechanisms is fundamental. This includes:

- Multi-factor Authentication (MFA): Requiring more than one method of authentication from independent categories of credentials to verify the user's identity.
- Role-Based Access Control (RBAC): Assigning system access to users based on their role within an organization, ensuring that individuals can only access information necessary for their duties.
- Least Privilege Principle: Granting users the minimum levels of access – or permissions—needed to perform their job functions.

7.4.2 Data Protection and Privacy

Protecting the integrity and confidentiality of data within IIoT systems is paramount, particularly given the sensitivity of the data involved. Essential measures include:

- Encryption: Applying strong encryption standards to data at rest and in transit to prevent unauthorized access and ensure data integrity.
- Data Masking: Obscuring specific data within a database to protect it from those who do not need to know.
- Regular Data Backups: Ensuring data resilience through regular backups that can be restored in the event of a cyberattack or data loss.

7.4.3 Network Security and Segmentation

Securing the network infrastructure that IIoT devices rely on is critical to preventing unauthorized access and ensuring the continuous availability of services. Key strategies include:

- Network Segmentation: Dividing the network into smaller, manageable segments or subnetworks to contain potential breaches and simplify security management.

- Firewalls and Intrusion Detection Systems (IDS): Deploying firewalls and IDS to monitor and control incoming and outgoing network traffic based on predetermined security rules.
- Secure Communication Protocols: Ensuring that all communications between IIoT devices and between devices and control systems use secure, encrypted protocols to prevent eavesdropping and tampering.

7.4.4 Device and System Integrity

Maintaining the integrity of IIoT devices and systems is essential for ensuring that they operate as intended, free from unauthorized modifications or tampering. This requires:

- Regular Software Updates and Patch Management: Keeping all software up-to-date with the latest patches to address vulnerabilities and enhance security.
- Secure Boot and Hardware Trust Anchors: Ensuring devices start in a known safe state and use hardware-based features to verify the integrity of their software and firmware.
- Tamper Detection: Implementing physical and logical mechanisms to detect and respond to tampering attempts on IIoT devices.

7.4.5 Incident Detection and Response

The ability to quickly detect and respond to security incidents is crucial for minimizing their impact. This involves:

- Continuous Monitoring: Implementing tools and processes for the continuous monitoring of IIoT systems to detect abnormal activities that may indicate a security incident.
- Anomaly Detection: Using advanced analytics and machine learning to identify patterns of behavior that deviate from the norm, potentially indicating a cybersecurity threat.
- Comprehensive Incident Response Plan: Developing and regularly updating an incident response plan that outlines procedures for responding to and recovering from security incidents.

7.4.6 Compliance with Regulatory Standards

IIoT systems often operate within regulatory environments that impose specific cybersecurity requirements. Ensuring compliance with relevant standards and regulations is essential for legal and operational reasons. This includes:

- Understanding Applicable Regulations: Identifying and understanding the regulations and standards that apply to the specific IIoT deployment, such as General Data Protection Regulation (GDPR), National Institute of Standard and Technology (NIST) CSF, or ISO/IEC 27001.
- Regular Compliance Assessments: Conduct regular assessments to ensure ongoing compliance with these standards and regulations, addressing any gaps as they arise.

7.4.7 Security Awareness and Training

Fostering a culture of security awareness among all stakeholders involved with IIoT systems is critical for preventing security breaches. Essential elements include:

- Regular Training Programs: Providing ongoing training to all employees on cybersecurity best practices, the specific risks associated with IIoT systems, and their roles in maintaining security.
- Phishing Awareness: Educating users on the dangers of phishing attacks and how to recognize and respond to them.

Securing IIoT systems requires a thorough and mixed approach that addresses a wide range of cybersecurity challenges. By implementing these key requirements, organizations can advance the security and resilience of their IIoT ecosystems, protecting critical industrial processes from cyber threats. As technology emerges and new vulnerabilities come out, organizations must remain aware, continuously updating and refining their cybersecurity practices to safeguard their IIoT deployments effectively.

7.5 Cyber Security Countermeasures for Industrial IoT

The IIoT has revolutionized industries by offering unparalleled levels of efficiency, productivity, and operational insights. However, the integration of digital and physical worlds also presents significant cybersecurity vulnerabilities that could potentially disrupt industrial operations and compromise sensitive data. Addressing these challenges necessitates a thorough suite of cybersecurity countermeasures tailored to the unique requirements of IIoT environments.

7.5.1 Network Segmentation and Isolation

Objective: To limit the attack surface and contain potential breaches within controlled zones.

- Implementation: Divide the network into segments based on functionality, sensitivity of the data handled, or the criticality of the operations managed. Use firewalls and gateways to regulate traffic between segments, applying strict rules that only allow necessary communications.
- Benefits: Enhances the overall security posture by minimizing the pathways accessible to attackers and reducing the impact of a compromise on the broader network.

7.5.2 Strong Authentication and Access Control

Objective: To ensure that only authorized users and devices can access IIoT systems and data.

- Implementation: Deploy multi-factor authentication (MFA) for user access, particularly for remote access and administrative accounts. Implement robust access control mechanisms for devices, such as certificate-based authentication or device fingerprinting.
- Benefits: Reduces the risk of unauthorized access due to compromised credentials or unauthorized devices attempting to connect to the network.

7.5.3 Encryption of Data in Transit and at Rest

Objective: To protect the confidentiality and integrity of data.

- Implementation: Utilize strong encryption protocols (such as Transport Layer Security (TLS) for data in transit and Advanced Encryption Standard (AES) for data at rest) to secure data at all stages. Ensure keys are managed securely, with periodic rotations and using hardware security modules (HSMs) where feasible.

- Benefits: Prevents unauthorized access to sensitive information, ensuring that data remains confidential and has not been tampered with.

7.5.4 Regular Software and Firmware Updates

Objective: To address known vulnerabilities in IIoT devices and software before they can be exploited by attackers.

- Implementation: Establish a patch management process that includes timely application of updates from device manufacturers and software vendors. For critical systems where immediate updates are not feasible, implement compensatory controls until patches can be applied.
- Benefits: Reduces the window of opportunity for attackers to exploit known vulnerabilities, thereby enhancing system security.

7.5.5 Continuous Monitoring and Anomaly Detection

Objective: To detect potential security incidents or system anomalies indicative of a cyberattack.

- Implementation: Deploy monitoring tools and IDS that can analyze network traffic, system logs, and device behavior in real time. Utilize advanced analytics and machine learning to identify patterns indicative of malicious activity.
- Benefits: Enables early detection of security threats, allowing for prompt response and mitigation to minimize potential damage.

7.5.6 Physical Security Measures

Objective: To prevent unauthorized physical access to IIoT devices and infrastructure.

- Implementation: Secure physical access to critical IIoT components through locks, surveillance cameras, and access control systems. Employ tamper-evident seals on devices to detect unauthorized physical modifications.
- Benefits: Protects against attacks that require physical access, such as tampering with devices to introduce malware or bypassing network security controls.

7.5.7 Incident Response and Recovery Plans

Objective: To ensure preparedness for, and effective management of, cybersecurity incidents.

- Implementation: Develop and regularly update an incident response plan that outlines roles, responsibilities, communication protocols, and procedures for addressing various types of security incidents. Conduct regular drills and simulations to test the plan's effectiveness.
- Benefits: Enhances organizational resilience by ensuring a coordinated and efficient response to cyber incidents, minimizing their impact, and facilitating rapid recovery.

7.5.8 Security Awareness and Training

Objective: To foster a culture of cybersecurity awareness among all stakeholders involved with IIoT systems.

- Implementation: Provide regular training sessions on cybersecurity best practices, specific risks associated with IIoT, and the importance of adhering to security policies. Include training on recognizing and responding to social engineering attacks.
- Benefits: Empowers employees to act as a first line of defense against cyber threats, reducing the likelihood of successful attacks due to human error.

Implementing these cybersecurity countermeasures is crucial for securing IIoT systems against the diverse risks and threats they face. By adopting a layered approach that addresses both technological and human factors, organizations can significantly enhance the resilience of their IIoT settings. Continuous evaluation and adaptation of these countermeasures are essential to staying ahead of evolving cyber threats and protecting critical industrial processes powered by IIoT technology.

Suggested Websites

- NIST Cybersecurity Framework (CSF) https://www.nist.gov/cyberframework

8

EFFECTIVE CYBERSECURITY COUNTERMEASURES FOR INDUSTRIAL IoT

8.1 Introduction to Effective Cybersecurity and Industrial Safety

In an era where the Industrial Internet of Things (IIoT) has become integral to the operations of critical infrastructure and manufacturing processes, the convergence of cybersecurity and industrial safety has emerged as a paramount concern. The digitalization of industrial systems, while bringing about unprecedented efficiency and productivity gains, also introduces new vulnerabilities and amplifies the potential impact of cyber threats on physical safety.

The IIoT ecosystem, characterized by its network of sensors, machines, and devices interconnected through the internet, has fundamentally transformed industrial operations. This digital transformation enables real-time monitoring, predictive maintenance, and optimize operational efficiency. However, it also creates a complex web of dependencies where cybersecurity vulnerabilities can translate directly into risks to physical safety and industrial processes. The integration of IT (information technology) with OT (operational technology) in IIoT environments has blurred the traditional boundaries between cyber and physical domains, making the alignment of cybersecurity and industrial safety strategies imperative.

As IIoT systems become increasingly critical to industrial operations, they become more attractive targets for cyber adversaries. These adversaries range from financially motivated attackers seeking to extort businesses through ransomware or data theft to state-sponsored actors aiming to disrupt critical infrastructure for geopolitical gain. The potential consequences of these cyber threats extend beyond data loss or operational downtime to include significant safety incidents,

DOI: 10.1201/9781003383253-8

environmental damage, and even loss of life. This evolving threat landscape necessitates a robust and proactive approach to cybersecurity tailored to the unique challenges of industrial environments.

Historically, cybersecurity and industrial safety have been treated as distinct disciplines, each with its methodologies, standards, and regulatory frameworks. However, in the context of IIoT, this siloed approach is no longer adequate. Effective cybersecurity measures are now a prerequisite for ensuring industrial safety, requiring a holistic strategy that integrates both domains. This integrated approach involves:

- Risk Assessment: Conduct comprehensive risk assessments that consider both cybersecurity threats and their potential impact on physical safety and industrial processes.
- Cross-Disciplinary Collaboration: Fostering collaboration between IT, OT, and safety teams to ensure a unified understanding of risks and the implementation of coordinated security and safety measures.
- Adherence to Standards and Best Practices: Aligning with industry standards and best practices that address both cybersecurity and safety considerations, such as the NIST Cybersecurity Framework, ISO/IEC 27000 series, and IEC 62443 for industrial cybersecurity.

This chapter aims to explore the critical components of industrial safety, examine how cybersecurity countermeasures can be aligned with safety objectives, and investigate emerging technologies that focus on enhancing industrial safety. By delving into these areas, we seek to provide a comprehensive understanding of how organizations can fortify their IIoT environments against cyber threats while ensuring the safety and reliability of their industrial operations.

In navigating the complexities of this intertwined landscape, the overarching goal is to establish resilient and secure industrial systems that not only protect against cyber threats but also uphold the highest standards of safety. As we venture deeper into this discussion, it becomes clear that achieving effective cybersecurity in the context of IIoT is not just about safeguarding information and systems—it is fundamentally about protecting people, processes, and the environment from harm.

8.2 Components of Industrial Safety

Industrial safety encompasses the measures, practices, and policies aimed at minimizing the risk of accidents, injuries, and fatalities within industrial settings, as well as protecting assets and ensuring continuous operations. In the context of the IIoT, where the digital and physical realms converge, industrial safety must also consider the cybersecurity vulnerabilities that could lead to physical harm.

8.2.1 Risk Management and Assessment

Risk management is the cornerstone of industrial safety, involving the identification, analysis, and mitigation of risks that could lead to accidents or incidents. In an IIoT-enabled industry, this process extends to cybersecurity risks that could have physical safety implications. A comprehensive risk management strategy includes:

- Hazard Identification: Systematically identifying potential hazards that could cause harm, including both physical hazards and cybersecurity threats.
- Risk Analysis: Evaluating the likelihood and potential impact of identified hazards, taking into account the vulnerability of systems and the existing controls.
- Risk Mitigation: Implementing measures to reduce risks to acceptable levels through engineering controls, administrative controls, cybersecurity measures, and protective equipment.

8.2.2 Physical Safety Measures

Physical safety measures are designed to prevent accidents and injuries from operational processes and machinery. These measures are particularly critical in manufacturing, chemical processing, and other industrial sectors. Key components include:

- Machine Guarding: Installing physical barriers to protect workers from moving parts, electrical hazards, and other dangerous aspects of machinery.

- Emergency Shutdown Systems: Implementing systems that can quickly shut down operations in response to hazardous conditions, minimizing the risk of accidents.
- Personal Protective Equipment (PPE): Providing workers with appropriate PPE, such as helmets, gloves, and safety glasses, tailored to the specific hazards they may encounter.

8.2.3 Environmental Controls

Maintaining a safe working environment is essential for industrial safety, involving the management of air quality, noise levels, and exposure to hazardous substances. Effective environmental controls help prevent long-term health issues and contribute to a safer workplace. Measures include:

- Ventilation Systems: Ensuring adequate ventilation to remove harmful vapors, dust, and other contaminants from work areas.
- Noise Reduction: Implementing noise control measures, such as sound-dampening materials and equipment design modifications, to protect hearing.
- Hazardous Material Management: Properly storing, handling, and disposing of hazardous materials to prevent exposure and environmental contamination.

8.2.4 Training and Awareness

Educating employees about safety practices, emergency procedures, and the specific hazards associated with their work is crucial for preventing accidents and ensuring a safe response to incidents. Training programs should be comprehensive and regularly updated to reflect new risks and technologies. Components include:

- Safety Inductions: Providing new employees with an overview of safety policies, emergency procedures, and their roles in maintaining a safe workplace.
- Skill-Specific Training: Offering training programs tailored to the specific tasks and equipment employees will be working with, including safe handling of IIoT devices.

- Emergency Drills: Conduct regular drills to ensure employees are prepared to respond effectively to emergencies, such as fires, chemical spills, or cybersecurity incidents with physical safety implications.

8.2.5 Cybersecurity Measures

In the realm of IIoT, cybersecurity measures become an integral component of industrial safety, protecting systems from attacks that could lead to physical harm. Essential cybersecurity measures include:

- Network Security: Implementing firewalls, intrusion detection systems, and secure communication protocols to protect IIoT networks from unauthorized access and attacks.
- Device Security: Ensuring IIoT devices are securely configured, regularly updated, and monitored for signs of compromise.
- Incident Response: Preparing to respond swiftly to cybersecurity incidents with plans that address both the cyber and physical aspects of industrial safety.

8.2.6 Regulatory Compliance and Standards Adherence

Complying with industry-specific safety regulations and standards is essential for legal and operational reasons. Regulatory frameworks often provide guidelines for managing safety risks, while adherence to standards like ISO 45001 (Occupational Health and Safety) and sector-specific safety standards can help organizations implement best practices and achieve continuous improvement in safety performance.

The components of industrial safety in an IIoT context are multifaceted, encompassing traditional physical safety measures, environmental controls, training and awareness programs, and modern cybersecurity practices. Together, these components form a comprehensive approach to safeguarding workers, assets, and operations in the digital industrial era. As IIoT technologies continue to evolve, so too must the strategies and measures employed to ensure industrial safety, requiring ongoing assessment, adaptation, and commitment to creating a culture of safety and security.

8.3 Alignment of Cybersecurity Countermeasures with Industrial Safety

The integration of the IIoT into core industrial processes has blurred the lines between cybersecurity and physical safety. Cybersecurity incidents now have the potential to cause physical harm, making the alignment of cybersecurity countermeasures with industrial safety practices a critical concern. This section delves into how cybersecurity countermeasures can be harmonized with industrial safety objectives, providing examples and detailing strategies for achieving a cohesive approach to safeguarding IIoT environments.

8.3.1 Understanding the Convergence

The convergence of cybersecurity and industrial safety in the IIoT context arises from the interconnected nature of these systems. A cyberattack targeting an IIoT device could lead to the loss of operational control, resulting in safety hazards such as equipment malfunctions, environmental damage, or even injury. For example, in 2010, the Stuxnet worm targeted Supervisory Control and Data Acquisition (SCADA) systems, causing physical damage to centrifuges in a nuclear facility. This incident highlighted the potential for cyberattacks to cross into the physical domain, underscoring the need for integrated safety and security measures.

8.3.2 Strategic Alignment of Countermeasures

The strategic alignment of cybersecurity countermeasures with industrial safety involves several key aspects:

- Risk Assessment Integration: Combining cybersecurity risk assessments with traditional safety risk assessments ensures that potential cyber-induced safety hazards are identified and mitigated. This integrated approach should consider both direct risks (e.g., unauthorized access to control systems) and indirect risks (e.g., data integrity issues leading to incorrect operational decisions).
- Unified Incident Response Plans: Developing incident response plans that address both cybersecurity incidents and their potential safety implications ensures a coordinated response

to incidents. For instance, if a cyberattack disrupts an IIoT-monitored safety system, the response plan should include procedures for both addressing the cyberattack and activating manual safety controls or backup systems.

- Cross-Disciplinary Teams: Establishing cross-disciplinary teams comprising members from IT, cybersecurity, OT, and safety departments fosters collaboration and ensures that countermeasures address both security and safety concerns. Regular meetings and joint exercises can help these teams develop a unified understanding of risks and responses.

8.3.3 Examples of Aligned Countermeasures

- Secure Remote Access: Remote access to industrial control systems is essential for monitoring and maintenance but also presents a significant cyber risk. Implementing Virtual Private Networks (VPNs) with strong encryption, Multi-factor Authentication (MFA), and rigorous access controls protects against unauthorized access while ensuring that operators can safely manage systems remotely.
- Segmentation of Control Networks: Network segmentation isolates critical control systems from the broader network, reducing the risk of a cyberattack spreading to safety-critical systems. For example, segregating networks handling volatile chemical processes from general IT networks minimizes the risk of a malware infection compromising safety controls.
- Real-time Monitoring and Anomaly Detection: Deploying real-time monitoring and anomaly detection tools on IIoT devices and networks can quickly identify potential cyber threats or operational anomalies. Integration with safety management systems ensures that any detected anomalies trigger safety protocols, such as shutting down equipment or isolating compromised systems.
- Safety-Centric Device Management: Ensuring that IIoT devices are securely configured, regularly updated, and physically secured prevents tampering and cyber exploits that could lead to safety incidents. An example is the use of secure boot

mechanisms and firmware signing to prevent unauthorized modifications to device firmware that could compromise operational safety.

8.3.4 Best Practices for Alignment

- Develop Shared Governance Models: Establishing shared governance models for cybersecurity and safety helps ensure coordinated policy development, risk management, and incident response.
- Continuous Training and Awareness: Providing continuous training that covers both cybersecurity and safety topics ensures that all personnel are aware of the potential cyber risks to safety and the measures they can take to mitigate these risks.
- Leverage Industry Standards and Frameworks: Adopting industry standards that address both cybersecurity and safety, such as ISA/IEC 62443 for industrial automation and control systems, provides a structured approach to managing these intertwined risks.

The alignment of cybersecurity countermeasures with industrial safety is a crucial aspect of securing IIoT environments. By integrating risk assessments, developing unified incident response plans, and fostering collaboration across disciplines, organizations can ensure that their cybersecurity practices bolster, rather than compromise, industrial safety. As IIoT technologies continue to evolve, maintaining this alignment will be essential for protecting against increasingly sophisticated cyber threats that also carry significant safety implications.

8.4 Emerging Technology in Focus on Industrial Safety

As the landscape of industrial operations evolves with the integration of the IIoT, emerging technologies play a pivotal role in enhancing industrial safety. These innovations offer new ways to predict, prevent, and respond to safety hazards, aligning closely with cybersecurity measures to protect against both digital and physical threats. This section delves into several emerging technologies that are reshaping the approach to industrial safety, providing detailed insights into their applications and potential future impacts through hypothetical case studies.

8.4.1 *Artificial Intelligence (AI) and Machine Learning (ML) in Predictive Maintenance*

AI and ML technologies are revolutionizing predictive maintenance by analyzing data from IIoT devices to predict equipment failures before they occur, preventing accidents and ensuring operational continuity. By continuously monitoring sensor data from machinery, AI algorithms can identify patterns or anomalies indicative of potential failures. For example, an ML model could detect unusual vibrations or temperatures in a piece of equipment, triggering maintenance actions to address the issue before it leads to a safety incident. In a future scenario, a chemical manufacturing plant utilizes AI-driven predictive maintenance to monitor pressure valves throughout the facility. The system detects a subtle but abnormal fluctuation in pressure readings from one valve, predicting a potential failure within the next 48 hours. Automated alerts prompt immediate inspection and maintenance, preventing a catastrophic explosion and ensuring the safety of the plant and its workers.

8.4.2 *Blockchain for Secure and Transparent Safety Compliance*

Blockchain technology offers a secure, immutable ledger for recording and verifying transactions, which can be applied to ensure transparency and integrity in safety compliance and certification processes. Using blockchain, organizations can create tamper-proof records of safety inspections, compliance certifications, and maintenance actions. This not only enhances trust in the safety processes but also streamlines audits and regulatory compliance by providing an indisputable record of actions taken. Imagine a future where an IIoT -enabled manufacturing facility employs a blockchain-based system to log all safety-related activities, from equipment inspections to employee safety training completions. During a regulatory audit, the company easily provides verifiable records of its compliance activities, stored on the blockchain, significantly reducing the time and effort required for the audit process while reinforcing its commitment to safety.

8.4.3 Augmented Reality (AR) for Safety Training and Operations

AR technology overlays digital information onto the physical world, offering innovative ways to conduct safety training and assist workers in real time during operations. AR can be used to create immersive training experiences that simulate real-world safety scenarios without exposing workers to actual risk. Additionally, AR can assist workers in identifying hazards and performing complex tasks by providing real-time information and guidance directly in their field of vision. Soon, an oil and gas company will implement an AR-based training program for emergency response drills. Trainees wearing AR glasses are immersed in a virtual environment simulating a gas leak scenario, where they practice identifying the leak source and executing response protocols. The hands-on experience, enhanced by realistic simulations, significantly improves the trainees' ability to respond effectively to real emergencies.

8.4.4 Internet of Things (IoT) for Environmental Monitoring

IIoT devices are increasingly used for continuous environmental monitoring, detecting hazardous conditions such as toxic gas leaks, excessive heat, or equipment malfunctions that could lead to safety incidents. Sensors deployed throughout an industrial facility can monitor for environmental hazards and automatically alert workers and management to potential safety issues. This real-time monitoring enables swift responses to mitigate risks and prevent accidents. A future scenario sees a large-scale mining operation equipped with an extensive network of IoT sensors monitoring air quality for dangerous levels of methane. Upon detecting a methane concentration nearing unsafe levels, the system automatically initiates ventilation adjustments and alerts workers to evacuate the affected areas, averting a potential explosion and safeguarding worker safety.

8.4.5 Digital Twins and Their Impact on Industrial Safety

Digital twins represent a groundbreaking technology in the realm of industrial operations, particularly within the IIoT. A digital twin is a virtual replica of a physical system, process, or product, enabling businesses to simulate, monitor, and analyze their operations in a virtual space. This technology not only offers profound insights into performance and potential issues but also significantly enhances industrial safety by enabling predictive maintenance, risk assessment, and incident prevention.

Digital twins allow for the real-time monitoring and simulation of industrial environments, making them invaluable tools for identifying safety hazards, predicting equipment failures, and optimizing operational processes. By mirroring every aspect of a physical system in a digital format, digital twins provides a comprehensive overview of an operation's status, including potential safety issues that may not be immediately apparent in the physical world. Key safety enhancements include:

- Predictive Maintenance: Digital twins can predict when equipment might fail or require maintenance, preventing accidents caused by equipment malfunction. By analyzing data trends over time, these virtual models can forecast potential issues before they occur, allowing for proactive maintenance and repairs.
- Safety Training and Simulation: Digital twins offers a safe environment to simulate various scenarios, including emergencies, without any risk to human life or physical assets. This capability is invaluable for training personnel, allowing them to experience and react to potential safety incidents in a controlled, virtual space.
- Hazard Identification and Risk Assessment: Through the continuous monitoring of operational data, digital twins can identify anomalies that may indicate safety hazards, such as unusual machine behavior or environmental conditions. This real-time analysis aids in conducting thorough risk assessments and implementing preventive measures.

Digital twins offers a powerful tool for enhancing industrial safety, providing unparalleled insights into operational processes and potential risks. As this technology continues to evolve, its integration with IIoT systems will become increasingly vital for predictive maintenance, risk assessment, and safety training. By leveraging the capabilities of digital twins, industries can not only optimize their operations but also significantly improve the safety and well-being of their personnel and assets, paving the way for a future where industrial accidents are drastically reduced through proactive, data-driven strategies.

Emerging technologies such as AI, blockchain, AR, and IoT are transforming the field of industrial safety, offering innovative solutions to anticipate, prevent, and respond to safety challenges in IIoT environments. As these technologies continue to evolve and integrate with existing safety and cybersecurity measures, they hold the potential to significantly enhance the safety, efficiency, and resilience of industrial operations. Through strategic implementation and ongoing development, these technological advancements promise to create safer and more secure industrial workplaces in the future.

Suggested Websites

- National Institute of Standards and Technology (NIST)—Cybersecurity Framework (CSF) https://www.nist.gov/cyberframework
- Industrial Internet Consortium (IIC) https://www.iiconsortium.org/
- Center for Internet Security (CIS) https://www.cisecurity.org/
- ISA/IEC 62443 Standards https://www.isa.org/standards-and-publications/isa-standards/isa-iec-62443-series-of-standards
- Cybersecurity and Infrastructure Security Agency (CISA) https://www.cisa.gov/
- Open Web Application Security Project (OWASP)—IoT and Industrial Security https://owasp.org/

9
Cybersecurity Risk Assessment Methods

9.1 Introduction

Cybersecurity risk assessment is a systematic process that helps organizations identify, evaluate, and prioritize potential threats to their information systems. In the context of the Industrial Internet of Things (IIoT), the complexity of these environments, with their interconnected and resource-constrained devices, poses unique challenges and increases the risk of cyber-attacks. This chapter provides a structured guide to understanding, implementing, and visualizing cybersecurity risk assessments tailored for IIoT and industrial settings, focusing on industry-standard threat modeling techniques and risk analysis methods.

9.2 Principles of a Risk Assessment Approach

Risk assessment in cybersecurity is based on basic principles designed to identify, assess, and address risks effectively. These principles make certain that organizations embrace a structured, well-versed approach to protect their assets against threats and cyber-attacks.

- Confidentiality, integrity, and availability (CIA): These principles make sure that data is accessed only by authorized users, rests unchanged except by authorized sources, and is accessible whenever is needed it.
  - Confidentiality sees to it that sensitive data is only accessible to authorized users, decreasing the risk of unauthorized access or breaches.
  - Integrity ensures that data remains right and unchanged except by authorized actions, preventing potential attacks

of tampering and corruption that could compromise operations.

- Availability checks that data and resources are accessible to authorized users when needed, sustaining operational continuity and decreasing downtime.
- Systematic approach to risk assessment includes three vital steps:
  - Threat identification contributes to identifying potential threats related to the organization or industry, including cyber-attacks, insider threats, and environmental risks.
  - Vulnerability assessment is about finding weaknesses within systems or processes that could expose the organization to these risks and threats.
  - Impact calculation appraises the potential outcome of each threat based on the likelihood and severity of its strike on the organization's infrastructure and operations.

  Guided by these steps, organizations can gain a comprehensive understanding of their risk landscape and prioritize risks based on their potential impact.
- Risk tolerance levels at organizations need to define a sustainable level of risk based on factors like industry standards, business objectives, and regulatory obligations that could be considered. Furthermore, industry standards vary for different industries that have unique security requirements (finance or healthcare, where data protection standards are sensitive). Security countermeasures should align with the organization's goals and objectives to protect without unreasonably limiting operational flexibility. Regulatory requirements with laws and industry regulations are vital to avoid legal consequences and maintain client trust.

  An altogether understanding of risk tolerance allows organizations to portion out resources effectively, ensuring that critical assets are protected from risks, threats, and potential cyber-attacks.

These principles jointly provide a concrete foundation for a risk assessment process and permit organizations to structurally address potential cybersecurity threats while weighing up security needs with operational order.

9.3 Mitigation of Cybersecurity Risk

Mitigation strategies in cybersecurity center on minimizing the likelihood and impact of potential threats and cyber-attacks. By embracing targeted tactics and methods, organizations can effectively reduce their risk exposure and reinforce their toughness against cyberattacks. This part of the section provides vital approaches to cybersecurity risk mitigation related to IIoT and industrial contexts. Risk reduction strategies target vulnerabilities and security gaps within an organization's systems to minimize the risk of occurring cyberattacks. The ways that organizations and industries can benefit are through patch management, encryption, and network segmentation. Regularly updating and patch will address software vulnerabilities that attackers target to exploit. Patch management makes sure that systems remain stable against threats, preventing potential breaches due to outdated software. Data encryption at rest or transfer protects sensitive information, making it unreadable to the naked eye of unauthorized users and minimizing the impact of data breaches and data theft. From the perspective of infrastructure, network segmentation into small and isolated sections builds obstacles for attackers. For example, critical information systems or IIoT systems can be separated from those with less secure network areas, limiting the spread of attacks of lateral movement.

Cost–benefit analysis (CBA) has shown an effective mechanism of risk mitigation, which involves balancing security investments against the potential costs of cyber incidents to occur in the future. Through risk prioritization assessment of the likelihood and impact of each risk, organizations can prioritize the most high-impact critical ones for immediate mitigation toward a solution. On the other hand, lower-priority risks will receive fewer resources, enabling the organization to focus on the most significant threats. CBA guides in the allocation of resources efficiently, making sure that security countermeasures offer significant risk reduction compared to their costs. This point of view halts spending too much on low-level risk areas and makes sure critical infrastructure receives proper protection.

Continuous monitoring of the latest trends and evolving cybersecurity attacks, especially in IIoT settings, is a key element worldwide. A proactive approach seems to be persuasive in most industry sectors, having real-time detection, proactive defense, adaptive response, and so on. This systematic approach provides steps ahead from that incident happening; but without a multi-layered defense approach, which would allow organizations to manage cybersecurity risks effectively, it is hard to achieve. By integrating risk reduction techniques, CBA, and continuous monitoring, organizations can build a solid defense that adjusts to evolving threats and cyber-attacks and ensures continuing protection.

9.4 Implementation for IIoT

Implementing cybersecurity risk assessments in the IIoT settings needs a specific approach because of the distinctive features of connected industrial devices. This part speaks to the challenges, tailored risk assessment processes, and the role of edge computing in allowing effective cybersecurity countermeasures for IIoT systems. Challenges and constraints remain within the IIoT settings, which include various devices and limits on processing that impact the implementation of traditional security practices. Numerous IIoT devices, such as sensors and low-power modules, pose limited processing power, memory, and battery life. These limitations cut down on the use of encryption or authentication protocols and needed lightweight security measures. IIoT devices often operate in challenging environments, including industrial plants and outdoor locations, where they face physical issues (temperature extremes, humidity, and physical interference). These circumstances demand tough, low-maintenance security solutions that are both stable and efficient. The IIoT landscape comprises an extensive range of device types, each with unique operational roles and connectivity needs. This diversity mixes up the application of steady security policies and device-specific security configurations. Additionally, let us present a potential practical cybersecurity application for the industry considering the challenges mentioned. For example, the implementation of resilient, lightweight security protocols with an edge-based device management system. To building a custom lightweight security protocol to utilize energy-efficient we can use algorithms like elliptic curve cryptography (ECC). ECC provides the feasible solution on less computational power demanded. We can deploy those things enclosed resistant to any environmental factors for IIoT devices, considering risks of tampering with or interference. A centralized edge-based management system through an edge gateway can enforce various security configurations based on roles and profiles of devices and risks. Integration of lightweight encryption, physical strength, and adaptable policy management through edge computing, advanced the cybersecurity for mixed IIoT environments.

The following part of this section introduces a Python code sample that exhibits an edge-based device management system utilizing lightweight encryption (ECC) and simple device-specific policies. This code is a simple structure that can be extended to guide more complex IIoT systems.

Before running this, you'll need to install the cryptography library:

pip install cryptography

The following Python code defines device policies for IIoT devices and builds an edge management system, including the encrypting and decrypting process; first the device registration to a specific policy, and generating an ECC key pair for each device; for clearness, encryption is simulated through signing but can be enhanced further for encrypting.

```python
from cryptography.hazmat.primitives.asymmetric import ec
from cryptography.hazmat.primitives import hashes
from cryptography.hazmat.primitives.asymmetric.utils import encode_dss_signature, decode_dss_signature
from cryptography.hazmat.backends import default_backend

# Define device policies for IIoT devices
device_policies = {
    "sensor": {"encryption": "ECC", "encryption_strength": "low", "resource_limit": "low"},
    "camera": {"encryption": "ECC", "encryption_strength": "high", "resource_limit": "medium"},
    "actuator": {"encryption": "ECC", "encryption_strength": "medium", "resource_limit": "low"}
}

# Edge management system
class EdgeDeviceManager:
    def __init__(self):
        self.devices = {}
```

```
def register_device(self, device_id, device_type):
    if device_type in device_policies:
        policy = device_policies[device_type]
        self.devices[device_id] = {
            "type": device_type,
            "policy": policy,
            "key_pair": ec.generate_private_key(ec.SECP256R1(),
                default_backend())
        }
        print(f"Device {device_id} registered with policy: {policy}")
    else:
        print(f"Device type {device_type} not supported.")

def encrypt_data(self, device_id, data):
    device = self.devices.get(device_id)
    if device and device["policy"]["encryption"] == "ECC":
        # Encrypt data using device's ECC private key (simulated here
            as signing)
        private_key = device["key_pair"]
        signature = private_key.sign(data.encode(), ec.ECDSA(hashes.
            SHA256()))
        r, s = decode_dss_signature(signature)
        print(f"Data encrypted for device {device_id} with ECC. Sig-
            nature: (r={r}, s={s})")
        return r, s
    else:
        print(f"Device {device_id} not found or does not support ECC.")
        return None

def decrypt_data(self, device_id, r, s, data):
    device = self.devices.get(device_id)
    if device and device["policy"]["encryption"] == "ECC":
        # Decrypt data by verifying signature (simulated as verifying)
        public_key = device["key_pair"].public_key()
        signature = encode_dss_signature(r, s)
        try:
            public_key.verify(signature, data.encode(), ec.ECDSA(hashes.
                SHA256()))
```

```
        print(f"Data verified successfully for device {device_id}.")
    except:
        print(f"Data verification failed for device {device_id}.")

# Example Usage
edge_manager = EdgeDeviceManager()

# Register IIoT devices with specific types
edge_manager.register_device("sensor_1", "sensor")
edge_manager.register_device("camera_1", "camera")
edge_manager.register_device("actuator_1", "actuator")

# Encrypt and decrypt data for a registered device
data = "Sensor reading: 123.45"
r, s = edge_manager.encrypt_data("sensor_1", data)

# Attempt to decrypt (verify) the data
if r and s:
  edge_manager.decrypt_data("sensor_1", r, s, data)
```

Effective risk assessment for IIoT needs custom-making based on device type, network function, and location within the operational infrastructure. For example, high-value devices on critical network infrastructure may need more strict evaluation in contrast to border devices with minimal impact on core operations. As IIoT networks increase and alter to changing demands, risk assessment processes must be adjustable to board additional devices and new risk factors. Edge computing plays an important role in allowing timely and effective cybersecurity risk assessment for IIoT.

9.5 Threat Modeling Methodologies

Threat modeling proactively allows organizations to identify and address potential security flaws during the design and development process. Organizations can implement targeted mitigations by spotting risks early, reducing the likelihood and effect of potential cyber-attacks before they reach production. This part investigates the importance of threat modeling, discerns it from vulnerability assessment, and provides direction on selecting the most suited methodology. Threat modeling provides a structured approach to assess system design from a security view, allowing teams to identify lacks that attackers might exploit. By inspecting potential threats during the design phase, organizations can address security issues before they happen to be vulnerabilities in deployed production systems. Conveying security flaws in advance is broadly more cost-effective than mitigating changes after deployment. Threat modeling allows organizations to carry through a secure design basis in advance, reducing costly retroactive fixes. Proactively addressing risks and threats enhances an organization's security posture, minimizing the risk of data theft, service disruptions, and other negative impacts related to cyber-attacks.

While threat identification and vulnerability assessment are important to a thorough security strategy, they focus on various aspects of security. The process of threat modeling identifies likely risks essential to system design, supporting an image of how an attacker can exploit vulnerabilities or logical errors. Threat modeling highlights hypothetical cyber-attacks and potential security lacks, assisting architects in building security around the system. Vulnerability assessment, on the contrary, is centered on identifying existing security weaknesses in a system or application that has already been developed or running. Practical vulnerability assessment is complementary to cybersecurity practices activity known as penetration testing. It engages in scanning for known vulnerabilities (unpatched software, misconfigurations) and evaluating them to determine the organization's exposure to risks and threats. So, threat modeling leads to secure design and architecture, while vulnerability assessment makes sure that systems remain secure as they advance and change.

Each threat modeling methodology provides distinctive strengths and is best suited to various scenarios based on more than one factor, including system complexity, types of threats, and available resources. For intricate systems that involve numerous components and interactions, a detailed methodology like spoofing, tampering, repudiation, information disclosure, denial of service, and elevation of privilege (STRIDE) might be proper. Depending on the organization, there are various threat profiles. For example, financial institutions may order data integrity and confidentiality, while industrial settings focus on availability. Methodologies like Process for Attack Simulation and Threat Analysis (PASTA), which highlights a risk-central way, are suitable for critical risk industry-related threats. The availability of resources like time, budget, and competency for threat modeling may order the choice of methodology.

By understanding the significance of threat modeling, the clear roles of threat identification and vulnerability assessment, and the features of methodology, organizations can select the best approach to proactively secure their systems. Implementing the upright threat modeling methodology allows organizations to build security around their systems from the beginning and ensure protection against a range of cyber threats.

9.6 Referenced Threat Modeling Types

The following threat modeling methodologies provide marked benefits for securing IIoT systems. Furthermore, ways to evaluate, prioritize, and mitigate potential cybersecurity risks.

9.7 Common Vulnerability Scoring System (CVSS)

CVSS gives a numerical score to vulnerabilities based on severity, benefiting organizations that prioritize risks. For IIoT, CVSS scores aid in assessing vulnerabilities over various devices and focusing resources on addressing the critical findings based on their exploitability and impact.

9.8 Attack Trees

Attack trees describe potential attack vectors in a hierarchical structure, with each node as a potential attack vector. In IIoT, these trees aid in visualizing how attackers exploit connected devices and network vectors, leading security teams to mitigate vulnerable areas and secure critical parts within the system.

9.9 Process for Attack Simulation and Threat Analysis (PASTA)

PASTA is a risk-centered methodology centering on real-world attack scenarios. In IIoT, PASTA aids simulate attacks that could target industrial systems, enabling security teams to view the impact from an attacker's outlook and prioritize countermeasures for critical-risk areas.

9.10 STRIDE

STRIDE categorizes six types of threats, including spoofing, tampering, repudiation, information disclosure, denial of service, and elevation of privilege. For IIoT, STRIDE offers a structured approach to analyze security over device functions and make sure each threat type is taken care of and mitigated to maintain the operational integrity of the infrastructure.

9.11 Quantitative Threat Modeling Method (qTMM)

qTMM offers quantifiable analyses of threat risks, making it especially effective for complex IIoT settings with different threat sources. By quantifying risk, qTMM aids industrial organizations in assessing and allocating resources based on accurate threat impact and likelihood metrics.

9.12 Cyber-Attack Trees

Cyber-attack trees provide a structured perspective of potential attack vectors, enabling organizations to analyze and fight ways an attacker compromises IIoT systems. Cyber-attack trees are mainly useful in IIoT, where interconnected devices present numerous entry points. By mapping common attack patterns, security teams can identify critical-risk areas within the IIoT network and order protection for the most critical vulnerable devices.

9.13 Visualizing Cybersecurity Risks

Visualizing cybersecurity risks plays a key role in IIoT settings, as it allows stakeholders to understand the threat outlook and make informed decisions rapidly. Effective visualizations make intricate data more accessible and actionable, particularly in systems where numerous devices and network layers contribute to comprehensive risk. Visual tools, including heat maps, graphs, and dashboards, make

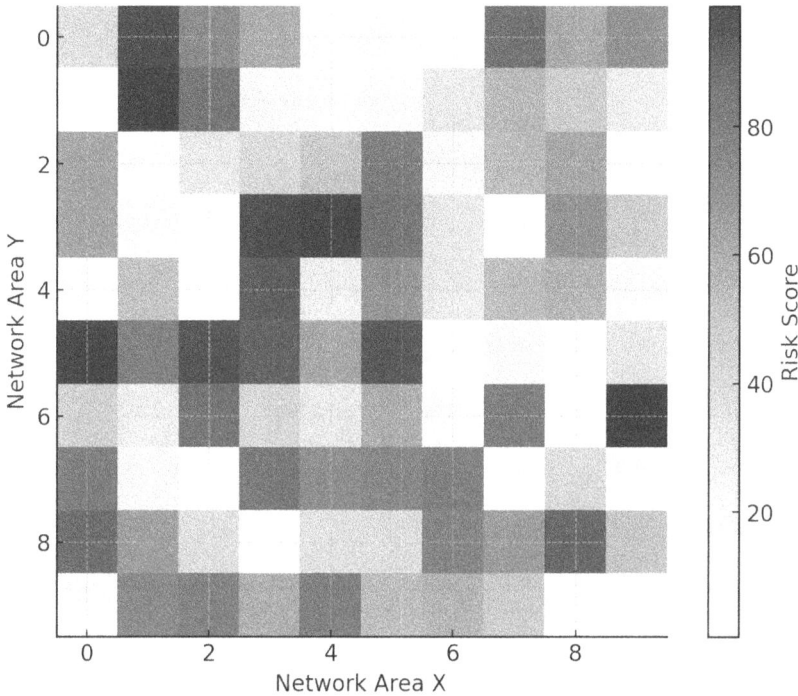

Figure 9.1   Risk heat map for IIoT network.

simpler intricate risk data, providing clarity and presentation for decision-makers. In IIoT, where various devices and protocols create complex networks, visualization allows quick identification and understanding of critical risks and supports communication over technical and non-technical audiences. Figure 9.1 generated from sample data utilized Python custom code identifying areas within the IIoT network that are more vulnerable to cyber threats.

Figure 9.2 presents a ranking of various devices by their risk score to order security efforts.

Figure 9.3 introduces an easy dashboard view of key metrics, such as incident alerts and average risk scores, proportion of high-risk versus low-risk devices.

Visualizations allow IIoT organizations to transmit risks effectively, maintain a view into security posture, and allocate resources to protect critical assets.

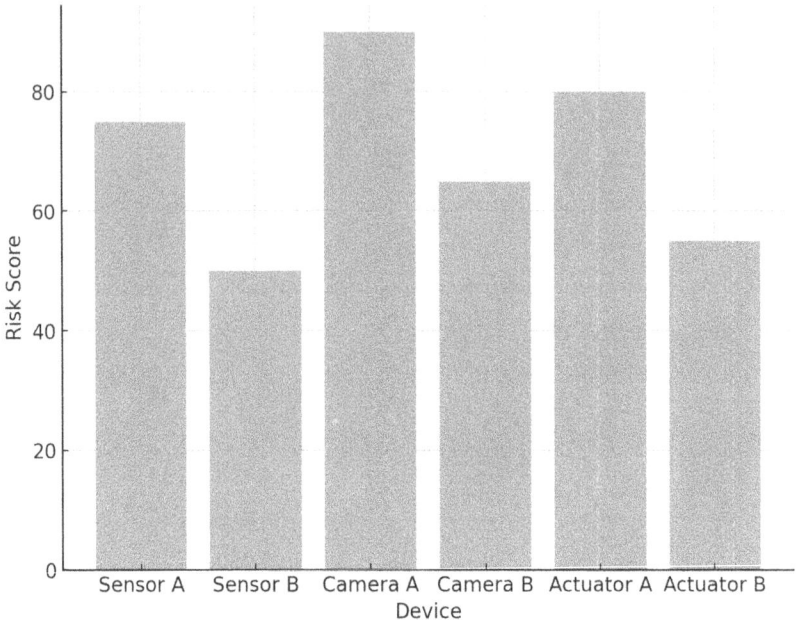

Figure 9.2 Risk scores by device.

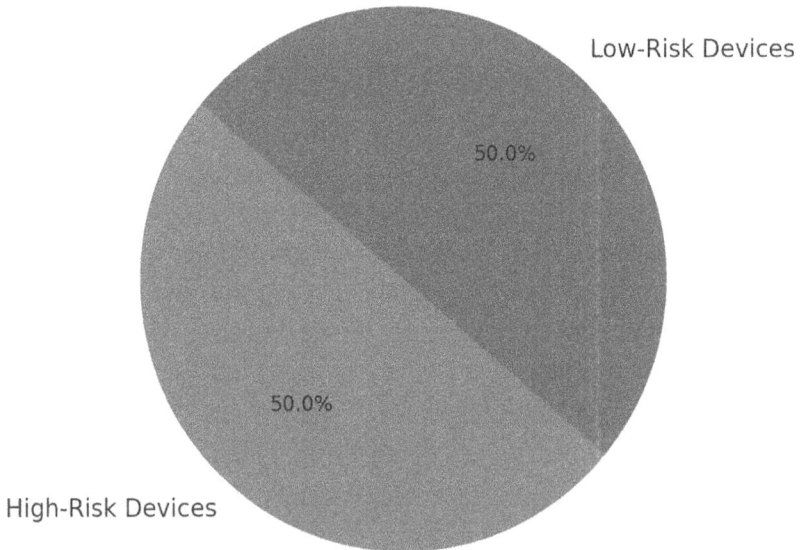

Figure 9.3 Device risk distribution.

Table 9.1 A Sample IIoT Cybersecurity Risk Analysis

THREAT MODEL	PURPOSE	APPLICATION IN IIOT	STEPS TAKEN IN RISK ANALYSIS	EXAMPLE OF DELIVERABLE
Common Vulnerability Scoring System (CVSS)	Prioritizes vulnerabilities based on severity and exploitability.	Aids IIoT managers assess and rank device vulnerabilities to allocate resources efficiently.	1. Identify device vulnerabilities. 2. Score each vulnerability using CVSS metrics. 3. Prioritize high-risk vulnerabilities for immediate action.	Vulnerability ranking report with CVSS scores for each device.
Attack trees	Maps potential attack paths hierarchically.	Visualizes potential attack paths on IIoT devices, helping to identify high-risk entry point.	1. Define possible attack goals (e.g., unauthorized access to control systems). 2. Create a tree diagram showing potential paths attackers could take. 3. Analyze the tree to identify critical paths and weak points.	Attack tree diagram showing critical pathways for each device type.
Process for Attack Simulation and Threat Analysis (PASTA)	Conducts risk-centric attack simulations.	Simulates specific attacks on IIoT systems, showing potential real-world impacts and helping prioritize countermeasures.	1. Define attack scenarios based on real-world threats (e.g., data tampering in transportation sensors). 2. Simulate each scenario to observe potential impacts. 3. Rank scenarios by impact and prioritize responses.	Risk assessment report with impact ratings and prioritized scenarios.
STRIDE	Identifies threats in six categories: spoofing, tampering, repudiation, information disclosure, denial of service, and elevation of privilege.	Analyzes IIoT devices for specific threat types across the network, ensuring that all security aspects are addressed.	1. Apply STRIDE categories to each device in the system. 2. Identify which threats are relevant for each device type. 3. Develop specific mitigation strategies for each identified threat.	STRIDE analysis matrix detailing threats and mitigations for each device.
Quantitative threat modeling method (qTMM)	Quantifies threat risks for measurable analysis.	Provides a structured, measurable approach for complex IIoT environments with varied threats and impacts.	1. Define potential threats and quantify the likelihood and impact for each. 2. Calculate overall risk scores. 3. Use quantitative scores to guide resource allocation for high-risk threats.	Quantitative risk report with risk scores and recommendations for each threat.

9.14 Risk Analysis of IIoT and Industrial Infrastructure

A comprehensive risk analysis for IIoT and the industrial environment must consider the general and sector-related factors. Through identifying and understanding vital risks, organizations can implement targeted measures that advance security and operational resilience. Internet of Things (IoT) settings face related risks due to their high interconnectivity and reliance on remote data gathering and transfer. Among many risks, we would consider unauthorized access, vulnerabilities in communication protocols (Message Queuing Telemetry Transport (MQTT), Constrained Application Protocol (CoAP)), and data integrity issues (tampering, theft). Sector-related risks can be diverse, depending on operational needs and infrastructure. In manufacturing, protecting production line machinery is vital, as any disruption (unauthorized access, tampering) can stop operations and lead to major financial losses. Transportation systems rely heavily on communication between vehicles, traffic lights, and central systems; any risks (malicious interventions, unauthorized control) that occur have real impacts on this industry. Table 9.1 is a sample IIoT cybersecurity risk analysis based on the threat modeling types mentioned earlier. This would serve as deliverables for industries to engage in risk analysis by aligning security strategies with identified threats and selecting proper mitigation steps.

Suggested Websites

- National Institute of Standards and Technology (NIST) Cybersecurity Framework
- MITRE ATT&CK Framework https://www.nist.gov/cyber framework
- Industrial Internet Consortium (IIC) Resources on IIoT Security https://www.iiconsortium.org/iisf/
- FIRST.org Threat Modelling https://www.first.org/global/ sigs/cti/curriculum/threat-modelling

10

IMPLEMENTATION OF THE IIoT CYBERSECURITY COUNTERMEASURES

10.1 Introduction

This chapter introduces a structured, practical way to implement cybersecurity countermeasures guided for Industrial Internet of Things (IIoT) systems. As IIoT settings bring special challenges—including diverse device types, high interconnectivity, and limited resources—the need for enhanced security strategies is vital. This chapter will guide readers through the essential steps for establishing a secure IIoT infrastructure, focusing on resource allocation, process maturity, and governance to consolidate an effective defense. Important aspects include preventive and responsive measures. Preventive measures, including encryption, network segmentation, and secure communication protocols, are proactive in reducing vulnerabilities earlier than those that can be exploited. Responsive measures act when a security incident occurs, such as incident handling, real-time monitoring, and immediate mitigation.

10.2 Organization of the Process

Organizing an orderly cybersecurity process for IIoT systems is important for allowing security measures to be well-defined, regularly applied, and effectively carried on. This part of the section offers a framework to assist organizations in establishing a robust process that goes well with cybersecurity engagements with organizational goals and needs. This structured cybersecurity process is the basis for effective IIoT cybersecurity and includes several steps. First, place clear security goals that go well with the organization's mission and wrap aspects of data protection, system availability, and compliance. Second, appoint roles and responsibilities within the cybersecurity team, making sure that each team member has a clear part in maintaining security. Lastly, prepare and draft standard operating procedures (SOPs) for cybersecurity process tasks, which standardizes all actions and benefits in maintaining the security posture of the organization.

The Control Objectives for Information and Related Technologies (COBIT) framework provides a well-structured way to align cybersecurity processes with organizational objectives from top to bottom. COBIT determines principles for risk assessment, management, and control, which are vital for IIoT settings with high risks. Sketch out control objectives that lead the consolidation of secure IIoT processes like data integrity, availability, and compliance management. Rich, full COBIT metrics can evaluate organizations' effectiveness of cybersecurity processes and areas for improvement. Complements of the COBIT is the library of ITIL, which offers detailed IT service management processes that assist cybersecurity within IT operations, Security Operation Center (SOC), and Computer Emergency Response Teams (CERT) teams. Well-defined planning and implementation come through project planning and implementation methodology, which are critical to maintaining a secure IIoT process.

10.3 Resources Aspects

Implementing adequate cybersecurity in IIoT settings needs a balanced investment in human and technical resources. Here would be needed an outline of the vital resource considerations, including supporting a robust cybersecurity framework. Crucial roles include cybersecurity analysts, IIoT network engineers, and incident response experts, responsible for managing various aspects of IIoT cybersecurity. Clear roles allow team members to understand their responsibilities and make sure of proper monitoring and threat detection. Technical resources would be necessary for IIoT cybersecurity, including intrusion detection systems (IDS), firewalls, endpoint protection, and network segmentation hardware.

An organized budget planning for cybersecurity allows for support of instant needs and future investments. Taking into consideration the cost of tools and hardware, maintenance, and upgrades, continuing investments support avoiding a lack of cybersecurity. Cybersecurity threats and cyber-attacks are evolving in techniques and methods, so training and staff awareness are vital. Personnel needs to have regular professional education and training plans; a wide awareness building on simple security practices, such as identifying phishing or maintaining secure passwords, to minimize human-related vulnerabilities. A resources-investing organization can establish a durable IIoT cybersecurity setting.

10.4 Maturity-Level Models

Maturity-level models offer a structured framework for evaluating and upgrading an organization's cybersecurity capabilities. By understanding their actual maturity level and aiming for specific upgrades, organizations can systematically advance their IIoT security posture. Capability maturity model (CMM) framework introduces an initial point to assess the organization's actual cybersecurity maturity. It specifies five levels of maturity, ranging from initial (ad hoc processes) to optimized (continual process improvement). For IIoT settings, the CMM can aid in identifying whether security processes are noteworthy and scalable over interconnected devices or if gaps exist that need immediate action. Furthermore, comprehensive process improvement (CMMI) enlarges the CMM framework by providing detailed support for advancing processes. In IIoT, this makes sure that security measures are not only settled but are also integrated into operational workflows.

Control Objectives for Information and Related Technologies (COBIT) maturity models assess the cybersecurity maturity-level ground on metrics, including governance effectiveness, process capability, and resource alignment. COBIT can establish alignment between IIoT cybersecurity processes and organizational goals, particularly when integrating security management with operational technology (OT). A significant milestone is reaching a level where incidents are identified, evaluated, and mitigated efficiently, minimizing disruption to IIoT operations. A benchmark of maturity is the setting up of official governance structures for cybersecurity, making sure of accountability, compliance, and strategic alignment to the organization. By using CMM, CMMI, and COBIT's maturity models, organizations can orderly evaluate and advance their IIoT cybersecurity processes, reaching milestones that advance security and operational resilience.

10.5 Optimized Process

Optimizing cybersecurity processes in IIoT settings is vital to ensure effective protection without unneeded intricacy or performance loss. We would explore strategies to well-run operations, implement continuous improvement, and balance security with system performance. The streamlining of cybersecurity operation will cover identifying and removing overlay or repeated tasks in security processes to increase efficiency, for example, monitoring tool security information and event management (SIEM). Automate repeated cybersecurity activities, including patch management, vulnerability scanning, and log analysis, to save time and reduce the likelihood of human error, e.g., AI-powered threat detection tools. SOPs are prepared and implemented for cybersecurity tasks to ensure consistency among teams and systems. Utilization of the data from previous incidents (lessons learned) guides to identify gaps in actual security measures and make needed adaptations. Regularly assess the effectiveness of cybersecurity countermeasures by risk assessments, pen testing, and employee training evaluations. Benefiting real-time monitoring tools can collect valuable insights during security operations and outcomes with actionable insights.

Design countermeasures protect devices and networks without notably downgrading their operational efficiency and implementing adaptive countermeasures that adjust based on network or device performance behavior to preserve a balance. Regularly tests IIoT systems on simulated operational loads to identify and mitigate performance bottlenecks led to by security processes. By streamlining operations, leveraging feedback loops, and balancing security with performance, organizations can create an optimized cybersecurity process for IIoT environments. This approach ensures robust protection while maintaining the efficiency and effectiveness of industrial operations. By optimizing operations, exploiting feedback loops, and balancing security with performance, organizations can create an optimized cybersecurity process for IIoT settings. This ensures stable protection while maintaining the efficiency and effectiveness of industrial operations infrastructure.

10.6 The Approach in Detail—Model

The model proposed in this book is based on three years of research conducted separately but provides a thorough and structured tool for advancing cybersecurity in IIoT infrastructures. It integrates four root functions to allow proactive and durable protection against cyber threats.

1. Penetration testing is vital in cybersecurity practice through the identification of vulnerabilities, applying black-box and white-box approaches, and utilization of various tools to scan, exploit, and document findings contribute to Penetration testing
2. Threat assessment and data grouping, analytics of findings from pen test on potential attack vectors and surfaces, a group based on category critical first (Open Worldwide Application Security Project (OWASP) methodology), and evaluating the impact of threats and prioritizing actionable steps.
3. Cost–benefit analysis (CBA) and countermeasure recommendation, assessing the financial aspect of applying the countermeasures, align the investments with organization plans and established standards (ISO).
4. Cybersecurity recommended countermeasures offer actionable mitigation steps to stakeholders and highlight countermeasures alignment with organizational goals.

The proposed model is designed to be stretchable and scalable, making it appropriate for various IIoT applications. This model finds applicability in various industry sectors such as manufacturing, smart cities, and energy sector. Implementing the proposed model starts with an initial evaluation to assess the actual cybersecurity posture of IIoT infrastructure. This step includes identifying critical resources, defining the scope of the cybersecurity endeavor, and defining clear objectives to assist upcoming actions. Once the vulnerabilities and risks were identified, the focus went to the implementation of countermeasures. Automated tools are integrated to support real-time monitoring and vulnerability scanning, while encryption protocols, access controls, and authentication mechanisms are implemented to secure devices and data transmission. These countermeasures allow IIoT settings to remain stable against known and emerging risks and threats. At last, continuous monitoring and reshaping are vital to maintaining an effective cybersecurity posture. By benefiting from real-time data from IIoT devices, organizations can detect anomalies, evaluate emerged risks, and instantly update countermeasures to address evolving vulnerabilities. This recurrent process ensures that countermeasures remain dynamic and aligned with the complex, transitional landscape of IIoT operations.

10.7 Inventory of Assets

An effective cybersecurity approach for IIoT settings begins with a thorough inventory of assets. This process enables all devices, systems, and data repositories to be considered for, classified, and monitored to protect against potential risks and threats. The first step in building a secure IIoT infrastructure is identifying and grouping critical assets based on their risk levels and significance to operational goals. Assets include sensors, actuators, gateways, network devices, and data repositories. An inventory management system is vital for keeping in sight all IIoT assets and maintaining their cybersecurity posture over time. To advance the security of an IIoT system, each asset must engage in a comprehensive vulnerability assessment to identify weaknesses and prioritize mitigation efforts.

For example, in a manufacturing premises, a real-time inventory of assets will identify controllers that manage production lines as critical. These controllers would be endlessly monitored for vulnerabilities and updated to prevent breaks. Likewise, in a smart city, traffic management sensors may be defined as high-risk due to their public-fronting nature and integral role in enabling safety. Proactive asset vulnerability evaluation and vigorous management systems allow these critical components to remain secure and working.

10.8 Governance and Implementation of Cybersecurity

Effective governance is important for implementing and maintaining a cybersecurity posture in IIoT settings. Governance structures allow cybersecurity activities to be coordinated; policies are enforced, and compliance with industry standards is reached. To supervise cybersecurity activities, organizations must consolidate governance structures that offer clear support and accountability. The main steps include a cybersecurity steering committee, risk management framework, and incident response team. As mentioned previously, the COBIT framework is recommended for IIoT governance. It offers an organized approach where you can align cybersecurity practices with organizational objectives. Additionally, NIST CSF supplements COBIT by centering on cybersecurity risk management and best practices through five functions: identify, protect, detect, respond, and recover. Apparently, defined roles are vital for cybersecurity governance, such as Chief Information Security Officer (CISO), Cybersecurity Analysts, incident response team, and device administrators. Cybersecurity policies are the principles of governance and must be aligned with industry standards and regulatory needs. By setting up robust governance assemblies, benefiting frameworks like COBIT and NIST CSF, and specifying clear roles and policies, organizations can effectively implement and maintain cybersecurity in IIoT settings. This way enables alignment with strategic goals, regulatory needs, and industry best practices.

10.9 Cybersecurity Countermeasures for IIoT

Effective cybersecurity countermeasures for IIoT infrastructures integrate preventive actions, stable controls, enhanced detection tools, and physical countermeasures. These grand designs work jointly to protect integrated devices, data, and networks from evolving risks and threats. Preventive measures focus on proactively minimizing vulnerabilities and protecting IIoT systems against cyber-attacks. Detection and response tools are vital for identifying and mitigating risks and threats in IIoT systems benefiting from SIEM systems, incident response measures, etc.

Securing and maintaining physical access to SIEM devices is crucial, particularly in settings where devices may be deployed in public or unmonitored areas. For example, in a smart manufacturing plant, countermeasures may include implementing firewalls to segment the production network infrastructure, using secure Message Queuing

Telemetry Transport (MQTT) protocols for real-time sensor communication, and implementing an intrusion detection system to monitor traffic for indication of intrusion. By integrating preventive measures, COBIT-guided controls, detection and response tools, and physical security control, organizations can enhance defense for IIoT systems. This thorough way addresses digital and physical vulnerabilities, enabling operational continuity and data security.

10.10 Functional Components

The functional components of a secure IIoT system are important to ensure cybersecurity countermeasures are effective and adaptable. These components address the main aspects of IIoT cybersecurity, from protecting data and devices to merging with existing systems. A secure IIoT system depends on vital components, including secure gateways, encryption tools, authentication mechanisms, and intrusion detection and prevention systems (IDPS). Secure gateways play as brokers, encrypting data and controlling device access. Encryption tools protect data at transit and rest, making sure of confidentiality and integrity (CI). Authentication mechanisms, including multi-factor authentication (MFA) and biometrics, guide to verifying user and device identities and reducing unauthorized access. These components jointly create a solid foundation for IIoT cybersecurity.

The COBIT framework offers an orderly approach to defining and aligning functional components with cybersecurity objectives. By highlighting the principles of confidentiality, integrity, and availability (CIA), COBIT allows component selection to aid these goals. It prioritizes resource allocation for critical components and positions them with organizational objectives. Also, COBIT's performance monitoring metrics allow continual evaluation and advancement, ensuring that cybersecurity components remain effective and appropriate over time.

Edge computing advances IIoT cybersecurity by delegating data processing and security functions. It minimizes latency by allowing local threat detection and response on edge devices, ensuring faster reaction times. Privacy is advanced as sensitive data is processed at the edge, reducing transferring to centralized servers and minimizing interception risks. Integrating new security components into actual IIoT infrastructure ensures the operational chain and advances security posture. This needs compatibility with actual protocols, such as MQTT or OPC UA, to keep interoperability.

10.11 Incident Handling

Incident handling is a vital element of IIoT cybersecurity, making sure that breaches are addressed quickly and effectively to reduce impact and prevent frequency repetition. A clear-defined incident response plan (IRP) is significant for managing security breaches in an organized and instant manner. The plan should provide clear steps for detection, containment, eradication, recovery, and reporting. It should include predetermined roles and responsibilities for incident response teams and set up communication protocols to bring together efforts over stakeholders. A thorough plan allows organizations to respond quickly, minimizing downtime and mitigating potential damage.

Early detection of incidents is vital and sensitive for minimizing their impact on the infrastructure. Techniques such as real-time monitoring with IDS and anomaly detection tools aid in identifying suspected activity in IIoT settings. Once an incident is identified, it must be escalated based on its severity and potential impact. A stepped escalation process makes sure that critical incidents receive first attention from response teams, while less critical issues are managed by operational staff members.

The recovery phase centers on restoring systems and operations to normal conditions after an incident. This includes eliminating malicious actors, patching vulnerabilities, and verifying system integrity. Post-incident analysis is essential, as it allows organizations to learn from the breach, know the root causes, and refine their cybersecurity readiness. Overseeing a comprehensive review and documenting lessons learned (LL) helps prevent similar incidents in the future and strengthens the general security posture.

Effective incident handling integrates proactive planning, swift detection and escalation, and tough recovery processes, enabling organizations to respond to risks and threats with agility and resilience.

10.12 Securing IIoT Devices from Cyber Threats and Cyber-Attacks

Securing IIoT devices is important to protect critical systems from threats and cyber-attacks. This includes implementing the measures to harden devices, secure networks, and manage vulnerabilities productively. To nourish IIoT devices, start with implementing secure configurations and minimizing attack surfaces. This means using secure firmware that aids encryption and authentication, disabling unused ports and services, and applying strong access controls. Systematic firmware updates and secure boot mechanisms make sure devices maintain their integrity and are immune to tampering or unauthorized modifications.

Network security is key for protecting IIoT devices and their data transmission. Network segmentation isolates sensitive devices and systems, preventing attackers from lateral movements within the network infrastructure. Traffic filtering, utilizing firewalls and IDS, monitors and blocks suspicious, malicious, or unauthorized communications. The benefits of encrypted communication protocols, such as Transport Layer Security (TLS) or algorithms, secure data transfer between devices and central systems.

Systematic vulnerability evaluation assists in identifying and addressing potential security lacks in IIoT devices. This includes scanning devices for known vulnerabilities, assessing the severity of risks, and prioritizing mitigation efforts. Administering timely patches and updates is vital to mitigate risks from newly identified exploits. Automated tools and centralized patch management systems smooth this process, making sure all devices remain secure and up to date.

By integrating device hardening, enhanced network security, and proactive vulnerability management, organizations can notably advance the security of their IIoT settings and protect them against cyber threats.

10.13 Cyber Security Lifecycle

The cybersecurity lifecycle offers an organized way to secure IIoT systems and ensure that security countermeasures are proactive, compliant, and continuously enhancing to address evolving cyber threats.

The cybersecurity lifecycle consists of the following stages:

- **Planning and Prevention**: Set up a solid security framework by identifying assets, evaluating risks, and defining policies for implementation. Implementation of preventive measures, including encryption, firewalls, and access controls, contributes to minimizing vulnerabilities.

- **Detection**: Monitoring systems in real-time utilization of IDS and anomaly detection tools results in identifying threats as they emerge.
- **Response**: Develop and execute IRP to handle and mitigate theft or breaches, reducing their impact on operations.
- **Recovery**: Restore the state to normal operations after an incident by repairing systems, applying patches, and ensuring that there are no remaining vulnerabilities.

The compelling nature of cyber threats needs a lifecycle approach that includes regular reviews and updates. Ongoing improvement ensures that emerged vulnerabilities are handled and lessons from past incidents are considered in further plans. Organizations should do periodic audits, simulate cyber threat scenarios, and refine security countermeasures to stay ahead of emerging risks. DevSecOps stands for development, security, and operations. Integrates security into every phase of the development and operations process and looks after a culture of shared responsibility for cybersecurity. In IIoT settings, this means built-in secure coding practices, automated vulnerability scanning, and regular security testing in the software development lifecycle. By ensuring security is an ongoing process, DevSecOps aids in identifying and addressing vulnerabilities in advance, reducing costs and risks for an organization.

By going after the cybersecurity lifecycle and embodying principles like continuous advancement and DevSecOps, organizations can build solid IIoT systems that are ready to counter evolving cyber threats while nurturing operational efficiency. This lifecycle way ensures that cybersecurity is still a dynamic and integral part of IIoT management.

Suggested Websites

- Center for Internet Security (CIS): Offers a range of guidelines, benchmarks, and security tools.
- SANS Institute: Provides cybersecurity training and resources tailored to IIoT and industrial settings.
- Industrial Internet Consortium (IIC): A valuable resource for standards and best practices in securing IIoT devices and networks.
- Proposed model tool: https://www.academyict.net/solutions/

Bibliography

Ansari S., Aslam T., Ansari A., Otero P., Ahmed I., and Maqbool F., "Internet of Things: Technologies and applications," *Introduction to Internet of Things in Management Science and Operations Research*, pp. 1–30, 2021, doi: 10.1007/978-3-030-74644-5_1

APT Groups and Operations. n.d. https://www.mandiant.com/resources/insights/apt-groups

Atzori L., Iera A., and Morabito G., "The Internet of Things: A survey," *Computer Networks*, vol. 54, no. 15, pp. 2787–2805, Oct. 2010, doi: 10.1016/j.comnet.2010.05.010

Atzori L., Lera A., and Morabito G., "The Internet of Things: A Survey," *Tạp chí Nghiên cứu dân tộc*, no. 24, Dec. 2018, doi: 10.25073/0866-773x/64

Bass B., Traditional security technologies and their role in modern cybersecurity. *Security Compass*, 2018. Retrieved from https://www.securitycompass.com/blog/traditional-security-technologies-and-their-role-in-modern-cybersecurity/

Behrendt A., de Boer E., Kasah T., Koerber B., Mohr N., and Richter G. Leveraging Industrial IoT and advanced technologies for digital transformation, 2021. https://www.mckinsey.com/~/media/mckinsey/business%20functions/mckinsey%20digital/our%20insights/a%20manufacturers%20guide%20to%20generating%20value%20at%20scale%20with%20iiot/leveraging-industrial-iot-and-advanced-technologies-for-digital-transformation.pdf

Behrendt A., de Boer E., Kasah T., Koerber B., Mohr N., and Richter G. A manufacturer's guide to scaling Industrial IoT, 2021. https://www.mckinsey.com/capabilities/mckinsey-digital/our-insights/a-manufacturers-guide-to-generating-value-at-scale-with-industrial-iot

Bhattacharjee S. and Nandi C., "Implementation of industrial Internet of Things in the renewable energy sector," *The Internet of Things in the Industrial Sector*, pp. 223–259, 2019, doi: 10.1007/978-3-030-24892-5_10

Blum D., *Rational Cybersecurity for Business*, USA: APress, 2020.

Boucher T. O. and Yalçin A., "Executing an information system design project," *Design of Industrial Information Systems*, pp. 261–291, 2006, doi: 10.1016/b978-012370492-4/50007-6

Boyes H., Hallaq B., Cunningham J., and Watson T., "The industrial internet of things (IIoT): An analysis framework," *Computers in Industry*, vol. 101, pp. 1–12, Oct. 2018, doi: 10.1016/j.compind.2018.04.015

Boyles S. D., Lownes N. E., and Unnikrishnan A., Transportation network analysis, 2019, https://sboyles.github.io/blubook.html

Britannica, "Information system," 2022. [Online]. Available: https://www.britannica.com/topic/information-system

Buja A., Apostolova M., and Luma A., "A model proposal for enhancing cyber security in industrial IoT environments," *Indonesian Journal of Electrical Engineering and Computer Science*, vol. 36, no. 1, p. 231, Oct. 2024, doi: 10.11591/ijeecs.v36.i1.pp231-241

Buja A., Apostolova M., Luma A., and Januzaj Y., "Cyber security standards for the industrial Internet of Things (IIoT)–A systematic review," *2022 International Congress on Human-Computer Interaction, Optimization and Robotic Applications (HORA)*, pp. 1–6, Jun. 2022, doi: 10.1109/hora 55278.2022.9799870

Buja A., Apostolova M., Luma A., and Januzaj Y., "Cyber security standards for the industrial Internet of Things (IIoT)–A systematic review," *2022 International Congress on Human-Computer Interaction, Optimization and Robotic Applications (HORA)*, pp. 1–6, Jun. 2022, doi: 10.1109/hora 55278.2022.9799870

Camarella S., Conway M. P., Goering K., and Huntington M. 2024. Digital twins: The next frontier of factory optimization. https://www.mckinsey.com/capabilities/operations/our-insights/digital-twins-the-next-frontier-of-factory-optimization

Caso J., Cole Z., Patel M., and Zhu W. Cybersecurity for the IoT: How trust can unlock value, 2023. https://www.mckinsey.com/industries/technology-media-and-telecommunications/our-insights/cybersecurity-for-the-iot-how-trust-can-unlock-value

Center for Internet Security. "Securing the Industrial Internet of Things (IIoT)", n.d. https://www.cisecurity.org/white-papers/securing-the-industrial-internet-of-things-iiot/

Chaudhary S., & Ahmed F. Ransomware: An overview of the threat landscape, 2020.

Cole E. "Advanced Persistent Threat: Understanding the Danger and How to Protect Your Organization, n.d.

Colina A. L., Vives A., Zennaro M., Bagula A., and Pietrosemoli E., *Internet of Things in 5 days*, 2016.

CompTia. "What is cybersecurity?," n.d. [Online]. Available: https://www. comptia.org/content/articles/what-is-cybersecurity

Cox J. Hackers took down part of the Ukrainian power grid. *Here's What We Know*, 2015, December 23). From https://www.washingtonpost.com/ news/the-switch/wp/2015/12/23/hackers-took-down-part-of-the-ukrainian-power-grid-heres-what-we-know/

Cybersecurity for Industrial Control Systems: SCADA, DCS, PLC, HMI, and SIS by David J. Stang (CRC Press, 2018.

Cybersecurity Ventures. The Colonial Pipeline ransomware attack: A case study on the financial impact of cyber-attacks, 2021. https://cyber securityventures.com/global-ransomware-damage-costs-predicted-to-reach-250-billion-usd-by-2031/

Davies K. *The VPNFilter malware: What it is and how to protect your network.* TechRadar, 2018.

Dhillon G. The role of traditional security technologies in modern cybersecurity. *FireEye*, 2018. Retrieved from https://www.fireeye.com/blog/products-and-services/2018/02/the-role-of-traditional-security-technologies-in-modern-cybersecurity.html

European Union Agency for Cybersecurity (ENISA). Ransomware attacks on the rise, 2020. https://www.enisa.europa.eu/publications/ransomware

Federal Bureau of Investigation. "Ransomware: What It Is and How to Protect Yourself," n.d. https://www.fbi.gov/scams-and-safety/common-scams-and-crimes/ransomware

Five Methodologies That Can Improve Your Penetration Testing ROI. n.d. https://www.eccouncil.org/cybersecurity-exchange/penetration-testing/ penetration-testing-methodology-improve-pen-testing-roi/

Gartner. "Gartner glossary," Gartner, n.d. [Online]. Available: https://www. gartner.com/en/information-technology/glossary/internet-of-things

Gilchrist A., "Introducing Industry 4.0," *Industry 4.0*, pp. 195–215, 2016, doi: 10.1007/978-1-4842-2047-4_13

Gubbi J., Buyya R., Marusic S., and Palaniswami M., "Internet of Things (IoT): A vision, architectural elements, and future directions," *Future Generation Computer Systems*, vol. 29, no. 7, pp. 1645–1660, Sep. 2013, doi: 10.1016/j. future.2013.01.010

Hanna A., Yoo J., Kim J., & Kim S. A survey of traditional information security technologies and their applications. *Journal of Information Security and Applications*, 44, 100847, 2019.

Henry J. and Lucas C., *Information Technology for Management*, Switzerland, 2009.

IIC, "IIRA," 2019. [Online]. Available: https://www.iiconsortium.org/pdf/ IIRA-v1.9.pdf

Íñiguez L. and Galar M., "A scalable and flexible open source big data architecture for small and medium-sized enterprises," *16th International Conference on Soft Computing Models in Industrial and Environmental Applications (SOCO 2021)*, pp. 273–282, Sep. 2021, doi: 10.1007/978-3-030-87869-6_26

Insights on the Internet of Things. n.d. https://www.mckinsey.com/featured-insights/internet-of-things/our-insights

ISACA. COBIT: Control objectives for information and related technology, 2021. Retrieved from https://www.isaca.org/cobit

ISO. ISO 27001:2013 - Information technology - Security techniques - Information security management systems – Requirements, 2021. Retrieved from https://www.iso.org/standard/64513.html

Izagirre U., Andonegui I., Landa-Torres I., and Zurutuza U., "A practical and synchronized data acquisition network architecture for industrial robot predictive maintenance in manufacturing assembly lines," *Robotics and Computer-Integrated Manufacturing*, vol. 74, p. 102287, Apr. 2022, doi: 10.1016/j.rcim.2021.102287

Jaloudi S., "Communication protocols of an industrial Internet of Things environment: A comparative study," *Future Internet*, vol. 11, no. 3, p. 66, Mar. 2019, doi: 10.3390/fi11030066

Jayalaxmi P., Saha R., Kumar G., Kumar N., and Kim T.-H., "A taxonomy of security issues in industrial Internet-of-Things: Scoping review for existing solutions, future implications, and research challenges," *IEEE Access*, vol. 9, pp. 25344–25359, 2021, doi: 10.1109/access.2021.3057766

Johnson J. Information security: The importance of protecting sensitive data. *TechTarget*, 2020. Retrieved from https://searchsecurity.techtarget.com/feature/Information-security-The-importance-of-protecting-sensitive-data

Jones K. J. Ransomware: A growing menace, 2018.

Kamara S. and Clarke K. L. Cryptovirology: Extortion-based security threats and countermeasures. In *Proceedings of the 13th ACM Conference on Computer and Communications Security (CCS '06). ACM*, 2006.

Kim J. WannaCry ransomware attack: Everything you need to know, 2017, May 12). From https://www.cnn.com/2017/05/12/tech/wannacry-ransomware-attack-explainer/index.html

Krebs B. Internet of Things gets its own Internet of Threats, 2016. From https://krebsonsecurity.com/2016/10/who-makes-the-iot-things-under-attack/

Lee E. A., Seshia S. A., and Jensen J. C., "Teaching embedded systems the Berkeley way," *Proceedings of the Workshop on Embedded and Cyber-Physical Systems Education*, pp. 1–8, Oct. 2012, doi: 10.1145/2530544.2530545

Lee N. K., Li X., and Wang D., "A comprehensive survey on genetic algorithms for DNA motif prediction," *Information Sciences*, vol. 466, pp. 25–43, Oct. 2018, doi: 10.1016/j.ins.2018.07.004

Lee Y. S. and Stankovic J. A. *Cyber-physical systems security: A research landscape.* ACM Computing Surveys, 2017.

Li J. and Jha N. *Smart Meter Security: Threats and Countermeasures.* IEEE Security & Privacy, 2012.

Madakam S. and Uchiya T., "Industrial Internet of Things (IIoT): Principles, processes and protocols," *The Internet of Things in the Industrial Sector*, pp. 35–53, 2019, doi: 10.1007/978-3-030-24892-5_2

McAfee. Advanced Persistent Threat (APT) - An Overview, n.d.

McKinsey Technology Trends Outlook. 2024. https://www.mckinsey.com/capabilities/mckinsey-digital/our-insights/the-top-trends-in-tech

Mehta B. R. and Reddy Y. J., "Industrial networks," *Industrial Process Automation Systems*, pp. 341–363, 2015, doi: 10.1016/b978-0-12-800939-0.00010-3

Michael Howard S. L., *The Security Development Lifecycle*, Mircosoft Learning, 2006.

Michels J. S., "Industrial connectivity and industrial analytics, core components of the factory of the future," *The Internet of Things*, pp. 247–270, Nov. 2017, doi: 10.1007/978-3-662-54904-9_15

Microsoft. "A guide to IoT technologies and protocols," n.d. [Online]. Available: https://azure.microsoft.com/en-us/solutions/iot/iot-technology-protocols/

Monteiro A. C. B. et al., "A look at IIoT: The perspective of IoT technology applied in the industrial field," *The Industrial Internet of Things (IIoT)*, pp. 1–29, Feb. 2022, doi: 10.1002/9781119769026.ch1

Munirathinam S., "Industry 4.0: Industrial Internet of Things (IIOT)," *The Digital Twin Paradigm for Smarter Systems and Environments: The Industry Use Cases*, pp. 129–164, 2020, doi: 10.1016/bs.adcom.2019.10.010

Nalini T. and Krishna T. M., "Analysis on security in IoT devices—An overview," *The Industrial Internet of Things (IIoT)*, pp. 31–57, Feb. 2022, doi: 10.1002/9781119769026.ch2

National Cyber Security Alliance. "Ransomware: A Growing Threat to Small Businesses," n.d.

National Institute of Standards and Technology (NIST) Cybersecurity Framework. n.d.

NIST. Glossary information technology (IT), NIST, n.d. [Online]. Available: https://csrc.nist.gov/glossary/term/information_technology

NVD NIST. n.d. https://nvd.nist.gov/vuln

OSINT Framework. n.d. https://osintframework.com/

Penetration Testing Methodologies. n.d. https://owasp.org/www-project-web-security-testing-guide/latest/3-The_OWASP_Testing_Framework/1-Penetration_Testing_Methodologies

Pereira C. E. and Neumann P., "*Industrial Communication Protocols*," Springer Handbook of Automation, pp. 981–999, 2009, doi: 10.1007/978-3-540-78831-7_56

Radware. IoT Botnets: The next wave of DDoS attacks, 2018. . From https://www.radware.com/getattachment/Security/Research/2025/Radware_ERT_Report_2017-2018_Final.pdf.aspx/?lang=en-US

Rathee G., Sharma A., Kumar R., and Iqbal R., "A secure communicating things network framework for industrial IoT using blockchain technology," *Ad Hoc Networks*, vol. 94, p. 101933, Nov. 2019, doi: 10.1016/j.adhoc.2019.101933

Rosenzweig P. *Cybersecurity and Cyberwar: What Everyone Needs to Know* (Oxford University Press, 2014.

Sabella R., Iovanna P., Bottari G., and Cavaliere F., "Optical transport for Industry 4.0 [Invited]," *Journal of Optical Communications and Networking*, vol. 12, no. 8, p. 264, Jul. 2020, doi: 10.1364/jocn.390701

Saranya V., Carmel Mary Belinda M. J., and Kanagachidambaresan G. R. "An evolution of innovations protocols and recent technology in industrial IoT," *Internet of Things for Industry 4.0*, pp. 161–175, Dec. 2019, doi: 10.1007/978-3-030-32530-5_11

Sari A., Lekidis A., and Butun I., "Industrial networks and IIoT: Now and future trends," *Industrial IoT*, pp. 3–55, 2020, doi: 10.1007/978-3-030-42500-5_1

Shunmughavel V., Challenges to industrial Internet of Things (IIoT) adoption, In *Innovations in the Industrial Internet of Things (IIoT) and Smart Factory* (pp. 117–132). IGI Global, 2021.

Sirotkin S., "5G Radio access network architecture: The dark side of 5G," Nov. 2020, doi: 10.1002/9781119550921

Sivaranjith, "Basics of industrial networking architecture," 2018. [Online]. Available: https://automationforum.co/basics-of-industrial-networking-architecture/

Smith J. *Essential Cyber Security*. Boston: Addison-Wesley Professional, 2019.

Stojanović M. D. and Boštjančič Rakas S. V., Eds., *Cyber security of industrial control systems in the future internet environment*. IGI Global, 2020. doi: 10.4018/978-1-7998-2910-2

Trend Micro. "Industrial Internet of Things (IIoT)," n.d. [Online]. Available: https://www.trendmicro.com/vinfo/us/security/definition/industrial-internet-of-things-iiot

Trend Micro USA. Advanced Persistent Threat (APT) – Definition, n.d.

Vmware. "What is network architecture?," Vmware, n.d. [Online]. Available: https://www.vmware.com/topics/glossary/content/network-architecture.html

Wang Y., Sun X., and Zhang J. Smart home security analysis and improvement. In *2017 IEEE International Conference on Internet of Things (iThings) and IEEE Green Computing and Communications (GreenCom) and IEEE Cyber, Physical and Social Computing (CPSCom) and IEEE Smart Data (SmartData)*, 2017.

Ward G. and Janczewski L., "Using knowledge synthesis to identify multidimensional risk factors in IoT assets," *Advances in Cyber Security*, pp. 176–197, 2021, doi: 10.1007/978-981-16-8059-5_11

Williams M. Why is IT security important? The importance of protecting sensitive data. *Cybersecurity Ventures*, 2020. Retrieved from https://cybersecurityventures.com/why-is-it-security-important-the-importance-of-protecting-sensitive-data/

Williams P. *Cyber Security: A Practical Guide*. Boston: Addison-Wesley Professional, 2020.

For Product Safety Concerns and Information please contact our EU
representative GPSR@taylorandfrancis.com
Taylor & Francis Verlag GmbH, Kaufingerstraße 24, 80331 München, Germany

FORENSIC
TOXICOLOGY
Medico-Legal Case Studies